EMERGENCY MANAGEMENT

MOBILE COMMAND & RESPONSE

VEHICLES

by

Anthony J. Rzucidlo

"A photographic review of emergency units"

Bloomington, IN  Milton Keynes, UK

authorHOUSE®

AuthorHouse™
1663 Liberty Drive, Suite 200
Bloomington, IN 47403
www.authorhouse.com
Phone: 1-800-839-8640

AuthorHouse™ UK Ltd.
500 Avebury Boulevard
Central Milton Keynes, MK9 2BE
www.authorhouse.co.uk
Phone: 08001974150

First published by AuthorHouse 10/26/2006

ISBN: 1-4259-4719-0 (sc)

Library of Congress Control Number: 2006906010

Printed in the United States of America
Bloomington, Indiana

This book is printed on acid-free paper.

<u>Reason for this Book</u>

My main reason for this publication was to share my interest in mobile command post vehicles with others with similar interest. Secondly, to showcase other types of emergency response support units utilized by "First Responders" from the public or private sectors.

My interest with mobile command post vehicles started in November 1993 when I completed my first 24-hour course in Incident Command, in my capacity as a security supervisor at the Ford Motor Company Research & Engineering Center in Dearborn, Michigan. At that time, I was responsible for fire, evacuation and training. Shortly after that the department received a 1994 Ford E-350 that was going to be used as a response/equipment unit for confined space rescue. It was my idea to also utilize it as a mobile command post unit for R&E Center Security. Needless to say, that I became directly involved with the vehicle as it applied to the maintenance of the unit and the equipment on it, in my new capacity as the special events, emergency preparedness and training supervisor for R&E Security until the fall of 2000.

<u>Special Thank You</u>

To my wife, Kimberly and my mother, Dolores Hnot for their support, advise, and assistance during the course of this project. Without them this book would of not have made it to print. Also to Don Nicholson for photographic assistance.

Book Format

The format for this book will contain background information on Incident Command, and the history and development of command post units. Also emergency response vehicles utilized by emergency responders will be covered.

The section dealing with mobile command units will consist of photographs of various command units mounted on commercial bodies costing hundreds of thousand of dollars, motor-homes, sport utility vehicles, trailers and vans.

Emergency response support vehicles will focus on the various types of specialized equipment utilized by "First Responders" and other emergency personnel to aid, and assist at the scene of a man made or natural disasters. These vehicles range from canteens, emergency supply equipment trailers, hazardous materials units, heavy rescues and lighting units just to mention a few. These pieces of specialized equipment can be found mounted on mini-buses, mobile homes, pick-up trucks, sport utility vehicles, trailers and vans.

Photos that appear in this publication either in the mobile command post section, or emergency response support vehicles section showcase vehicles from fire departments, emergency management agencies, emergency support activities, law enforcement agencies and private industry.

Background Information on the
Incident Command System

Perhaps some background information about the Incident Command System now referred to as the Incident Management System is in order to make the connection between mobile command units and the Incident Commander. As a result of the continuing problem with wild land fires in southern California in the 1970's, governmental agencies responsible for response to these incidents developed an organization known as "Firefighting Resources of California Organized for Potential Emergencies" created the Incident Command System. Even though the early formation of the Incident Command System (ICS) was primarily developed for wild land fire operations, it has developed into a management system that can be used for any type of emergency situation that requires command and control of resources be it equipment or personnel. It also can be utilized for non-emergency activities such as special events.

Over the years the Incident Command System and now in some circles referred to as the Incident Management System (IMS) has been accepted by law enforcement agencies, the private sector and other response groups. In the United States, the Federal government under federal law (Code of Federal Regulations) mandates the use of Incident Command in specific areas of emergency situations. Various forms of Incident Command have been established in other countries, which include, but are not limited to Australia, Canada, New Zealand and the United Kingdom of Great Britain.

As a result of the September 11, 2001 terrorist attack upon the United States, President George W. Bush issued a Homeland Security Presidential Directive that created a National Incident Management System. "NIMS" will provide a consistent nationwide approach for federal, state, and local governments to work effectively and efficiently together for, respond to, and recover from domestic incidents, regardless of cause, size, or complexity for interoperability and compatibility among federal, state and local capabilities, "NIMS" with a core set of concepts, principles, terminology, and technologies covering the incident command multi-agency coordination systems; unified command; training; identification and management resources (including systems for classifying types of resources); qualifications and certification collection, tracking, and reporting of incident information and incident resources.

Command Post History

In my opinion I would be willing to say going back some 25 to 30 years ago, and perhaps in some cases even longer that the agencies that now have some type of dedicated mobile command post did not have one then.

Some of the reasons might be that they (emergency management agencies, fire and police departments) did not feel the need for such a vehicle or that they might of fallen under a county plan which afforded them a mobile command post in the event that one was required.

On the other hand, during this same time period as noted above, that there were some agencies and units of government, especially during the Civil Defense era that had some type of dedicated mobile command unit, for the most part mounted on commercial type vehicles. I would venture to say that units during this time frame mainly would have be found in major cities, county departments as well on the state level.

In today's society the need for governmental units to be proactive as it relates to emergency management and preparedness and in light of the events of September 11, 2001, the need for a community to have a vehicle specially equipped as a mobile command post is vital to respond to man made or natural disasters.

Another reason for a community to have its own unit would be in the event of a major catastrophe covering numerous governmental units in a particular county, the availability of the counties command post might not be available.

Some jurisdictions have the ability to provide separate units to there fire and police departments, while others have both public safety agencies (fire/police) share one unit. Whatever the case, in my opinion one vehicle for both departments to utilize is better then none at all.

The days of working off the hood of a car or the back of a station-wagon or sport utility vehicle in all types of environmental (weather) related conditions still continues to some extent today for initial responses by Incident Commanders. However, for extended operations more likely then not, a dedicated mobile command post would be utilized.

Some industries are more proactive then others in the area of emergency management and preparedness. Some of them have identified the need for mobile command units and has either purchased new units or converted vehicles that they have already in their fleets to serve as command units.

In some cases, a company with several sites in close proximity to each other might decide to purchase one unit and locate it in a central area so that it would respond to a site without great delay, rather then duplicate a unit for each location, thus not creating a financial burden upon the organization for such items as equipment and maintenance.

In my opinion if a good business case has been made for the need for a mobile command post then it should be purchased. Private industry cannot always count on an immediate response from the public sector if an incident has struck an entire community where the business is located. Thus, you will need to be ready to manage the incident until public safety agencies can respond. The timeline in response could be a few hours to a few days in the worst case scenario depending on the incident.

Modern Era

So, as you view the various photos you will quickly see how over the course of the years these vehicles inmost cases developed into some high-tech command centers. A good example of this would be two units recently placed into service by the Nevada Department of Public Safety. The "Mobile Command Center's" are equipped with the following capabilities: Four positions for dispatch communications, voice over IP radio system, digital video recording system can capture all activities in the "MCC" along with what is on the 42-foot mast camera and the exterior of the coach. Eight-person conference room table along with bench seating for eight additional personnel. The driver's seat as well as the passenger's front seat can be utilized for additional seating, making the total amount of persons that can be seated within the conference room area 18. An electronic white board is used to capture anything briefed in the coach and also can send e-mail briefings out via satellite data transmission.

Video, television and power point presentations can be made to an outside audience as well as in the conference room area. Outside workstations have multiple phone and computer network connections to set-up a fully functioning office if needed outside under the exterior awning. In an underneath compartment, which slides out a microwave, refrigerator, coffeepot and outside water access can be found. An additional storage area is located in the "underbelly" of this unit which folding tables, chairs, scene lighting on tripods can be found.

Observations

Who knows what the future will hold as it applies for the development of mobile command units. I am quite sure that two factors that will help to influence future developments will be technology and the costs associated with purchasing these units. As long as governmental units are able to apply for and receive grants from the United States Department of Homeland Security for the purchase of mobile command units, I am sure that you will continue to see them placing customized units into service.

Governmental units that are not successful in obtaining grants for the purchase of a new command unit will continue to make the tough decisions based upon funding on what type of vehicle to obtain. The same could be said about those in private industry that are unable to obtain grants, and have to depend on internal funds from the interest that they represent.

In my opinion and based upon my involvement in a research project in obtaining a mobile command post, I would say that the two more favorable type of vehicles to be utilized for this purpose are a customized unit, motor-coach (bus) chassis and a mini-bus (commuter/transit) chassis style. The reasons for this would be first and most important that these units provide adequate room for personnel assigned to work in them. These vehicles afford the room to install the necessary support equipment needed for a command post to have in-order to deal effectively with an emergency incident. In addition they are able to include in most cases rest-room capabilities as well as offer an area for a conference room.

Perhaps costly by design, these vehicles can be directly driven to the scene of an incident and command and communications operations can start at once. Also if necessary the vehicle can be re-located at a moments notice.

Emergency response activities in the public sector and private industry may decide based upon several different reasons to utilize another type of vehicle to serve as a mobile command post. These vehicles can range from surplus government equipment, sport utility vehicles, trailers and vans. In my opinion each of the various type of vehicles mentioned above would function as a mobile command post, but have limitations in one form or the other.

In some cases government surplus vehicles (normally former military ambulances) may work out quite well for the needs of some communities. However, one must keep in mind that these vehicles were not originally designed to serve as command units, and more likely will require some modifications and in some cases drastic modifications to meet the needs of the agency that will be utilizing the unit. If these modifications cannot be done in-house or have the cost of the modifications donated by local businesses, it could end up being both costly and time consuming to the agency. Another concern for using this type of vehicle would be workspace and the ability to add necessary support equipment.

Even though pick-up trucks, sport utility vehicles and vans can be converted to serve the needs of a mobile command post some of the draw backs of these vehicles are a limited workspace for the Incident Commander and support personnel and the placement of support equipment. Needless to say no conference room can be provided for key personnel to meet during the time of an emergency. However, I would note that van chassis that are affixed to ambulance bodies afford a better working type unit over a van.

However, before we leave this section of pickup trucks, sport utility vehicles and vans, I will say that these units can serve as initial response units and could be used for minor incidents. Several agencies in the interest of getting the most for their money combine these types of vehicles to serve dual roles. These dual roles could range from air supply units, lighting units to rescues. However, vehicles that serve a dual purpose may not be as effective as they could be in the event of an incident based upon the fact that they are trying to perform more then one task.

Trailers being used, as mobile command units in my opinion are the most costly investment that either a public or private response activity can make. Factors that lead me in this view are not only do you have to use capitol to purchase the trailer, but you also need to purchase another vehicle if one is already not available to pull the trailer.

In addition to the double cost involved with a tractor/trailer unit as opposed to a single purpose unit, other considerations come into play. In incidents that I have seen a trailer deployed at, the unit that was utilized to bring the trailer to the scene was un-hooked and either parked else where or retuned back to its base of operation. If the need arose that the trailer needed to be re-located, time would be wasted waiting for the tow vehicle to return, or if on sight to hook-up to the trailer and move it.

The trailers that I have seen that are set-up as command posts do not seem to have the work areas that others vehicles afford. They seem to be chopped up as it relates to space. I am sure that their are some agencies that use trailers for command post purposes and that they work well for them, and that they are designed to afford proper workspace areas. However, I think that overall trailers are not the way to go, but the decision to purchase a trailer or not will rest with the individual activity.

Ultimately the final decision on what type of vehicle that will be purchased to be used as a mobile command post will be based upon the needs of the activity and funds available to be used towards the purchase. Whatever, type of vehicle is selected attempts should be made to adequately equip the unit with the tools be it computers, radios or workstations that will aid the Incident Commander in dealing with the emergency at hand. Also some thought should be given to future items that maybe purchased and installed on the unit. Its better to have the space available then try to create space or try and make it fit somehow.

<u>Emergency Response Support Vehicles</u>

Even though most of my focus has been on mobile command post units, emergency response support units are just as important and vital to the effective outcome of an incident. As you will see in this publication just like command units, these vehicles also range in different sizes and functions. However, each has their specific purpose that will aid the Incident Commander in dealing with whatever incident that they maybe faced with.

Mobile Command Post Photos

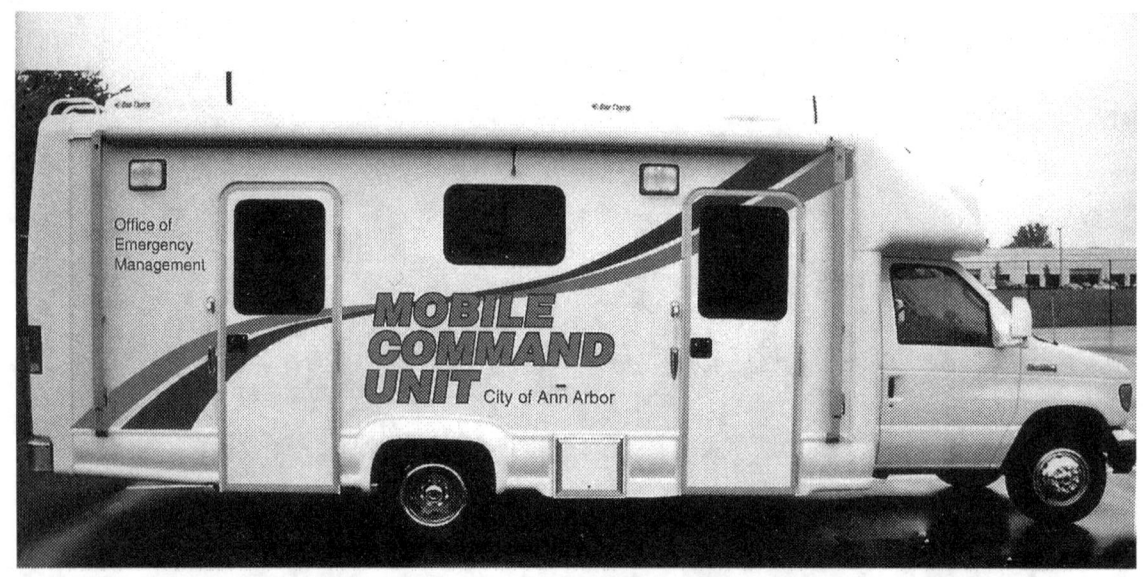

Ann Arbor (Michigan) Office of Emergency Managment mounted this command unit on a Ford chassis. The unit is painted white with grey and green lettering and accent stripping. Photo courtesy Ann Arbor Office of Emergency Managment.

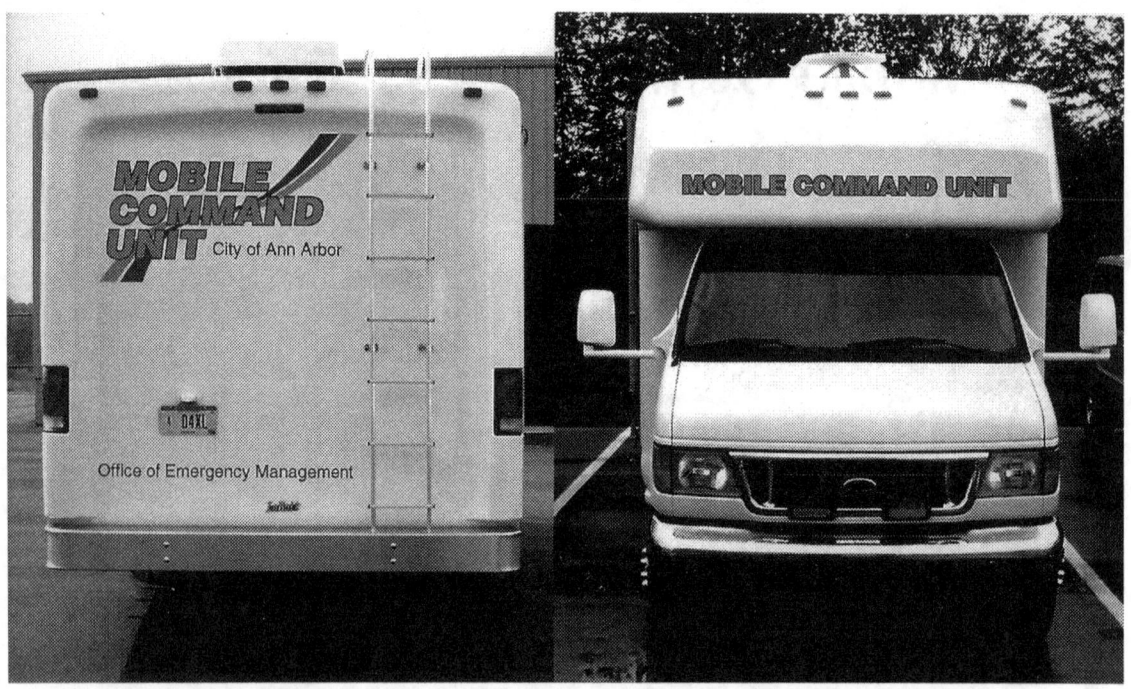

A somewhat different view of the Ann Arbor mobile command unit is presented, this time showing the unit in a rear and front photo shoot. Photo courtesy Ann Arbor Office of Emergency Managment.

Ajax Fire & Emergency Services (Canada) placed this 2006 Chevy Tahoe 4x4 into service as a command unit for the on duty "Platoon Chief" and carries the radio identification of "Car 5". Prior to this unit being placed into service, a van was utilized. Photo courtesy of Ajax Fire and Emergency Services.

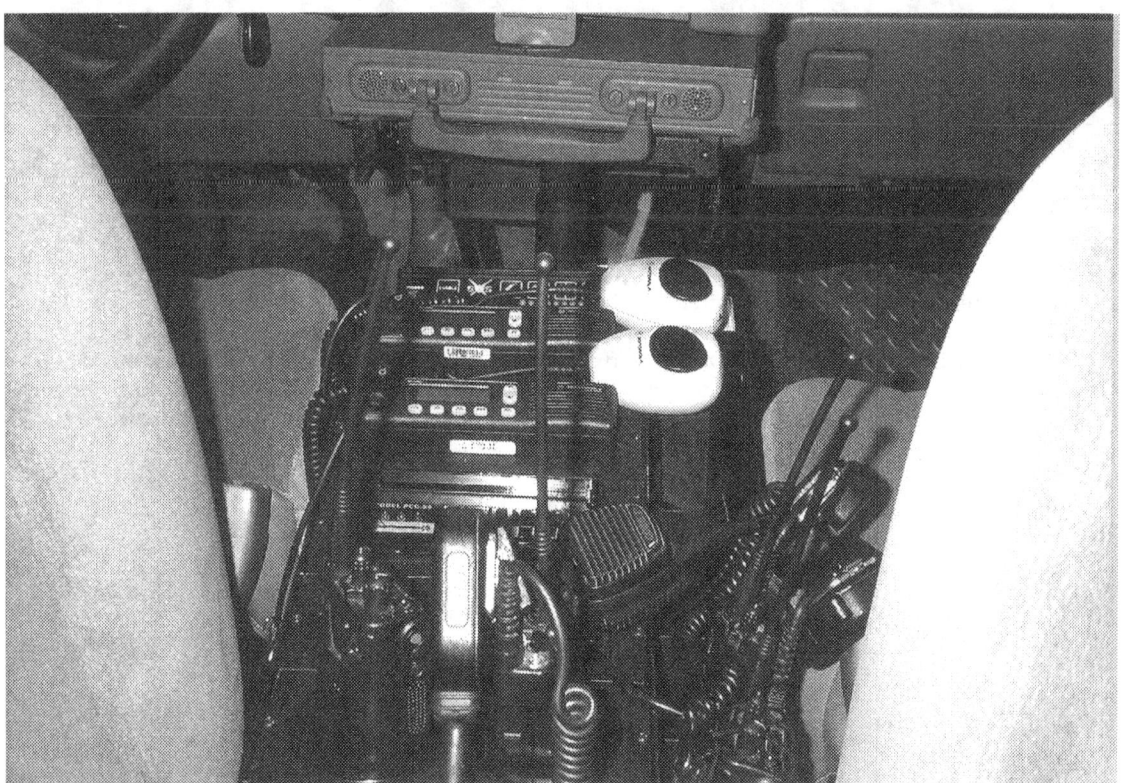

An interior shot of the new Ajax Fire & Emergency Services vehicle shows the set-up of the various radios and other emergency equipment. Photo courtesy of Ajax Fire & Emergency Services.

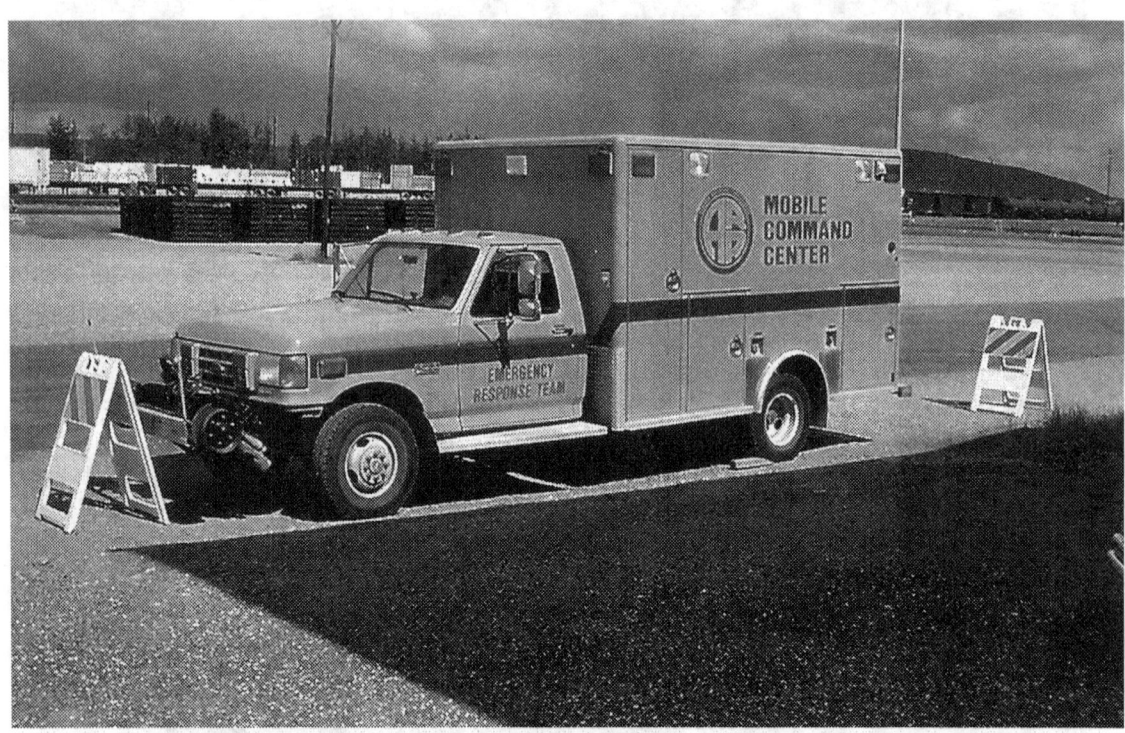

Alaska Railroad Corporation converted this 1991 Ford chassis former ambulance into a mobile command center for its Environmental Services & Response operations. Intersting feature about this unit is the hi-rail capability of this unit, meaning that it can travel on the roadway or on the rails depending on the incident. This unit is still in service in 2006. Photo courtesy of the Alaska Railroad Corporation.

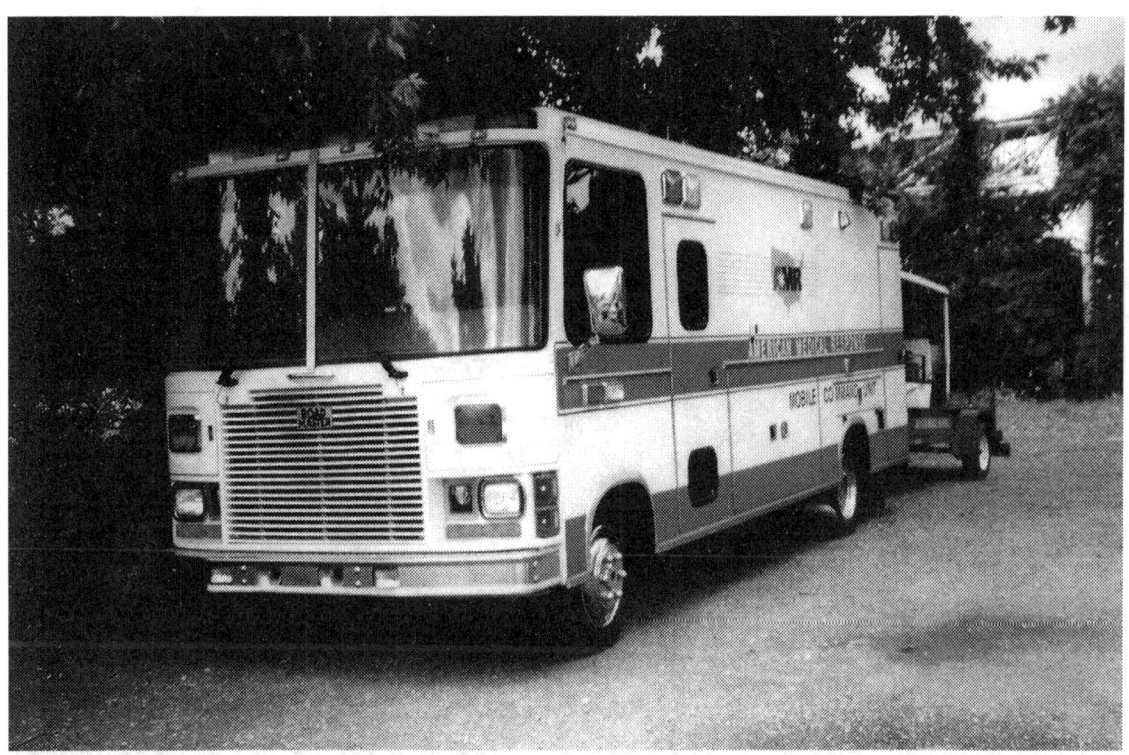

This medical command unit mounted on a Road Rescue chassis was operated by American Medical Response a nationwide ambulance operation. This unit with attached trailer was in operation in Michigan. Photo by Anthony J. Rzucidlo.

Bedfordshire Police (United Kingdom of Great Britain) operates this 2000 Mercedes as a command and control unit. Note the unique markings on the side of the vehicle, which is common in the UK. Photo by Christopher M. Taylor.

The British Transport Police (United Kingdom of Great Britian) utilized this major incident support trailer for various incidents within the areas of their jurisdiction. High visibility markings have been applied to the rear of the trailer. Photo courtesy British Transport Police.

California Emergency Mobile Patrol Search / Rescue an all volunteer / non profit organization utilizes this former Department of Water & Power 1982 GMC / Uniaden as a command unit. When not in service, the unit is stored at an LAPD station. Photo courtesy California Emergency Mobile Patrol.

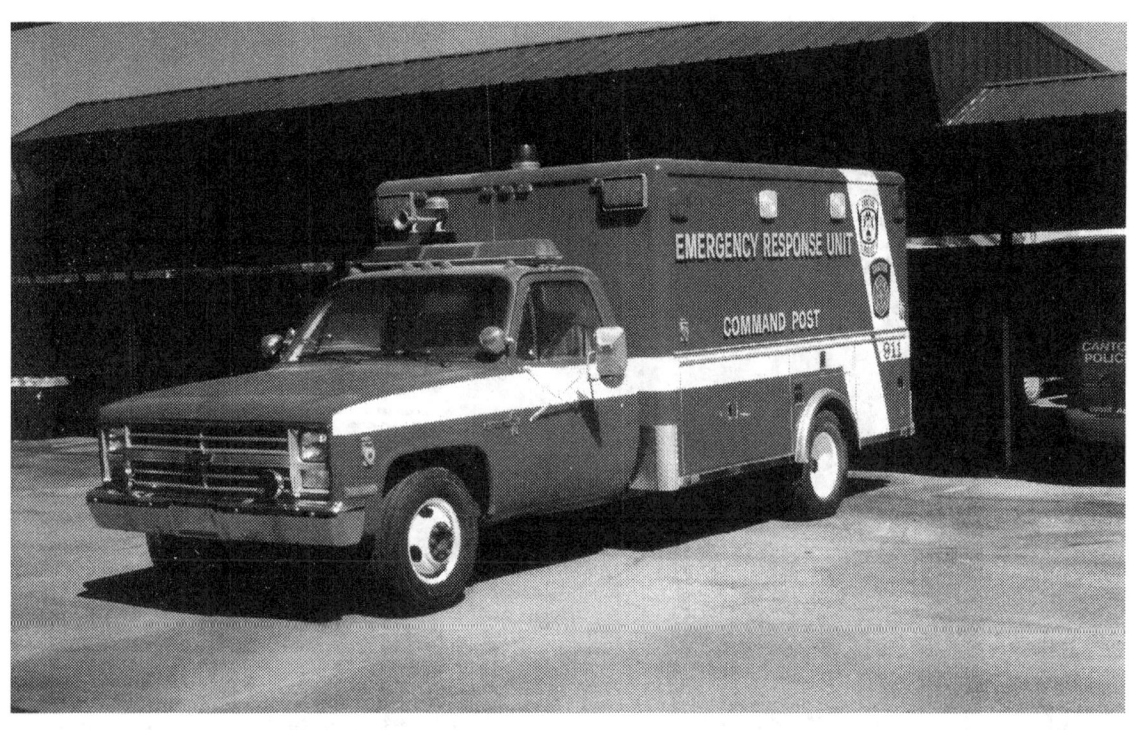

This 1987 Chevy fire rescue was converted into the very first mobile command post by Canton Township (Michigan). Both the fire department as well as the police department utilized this unit. Photo by Anthony J. Rzucidlo.

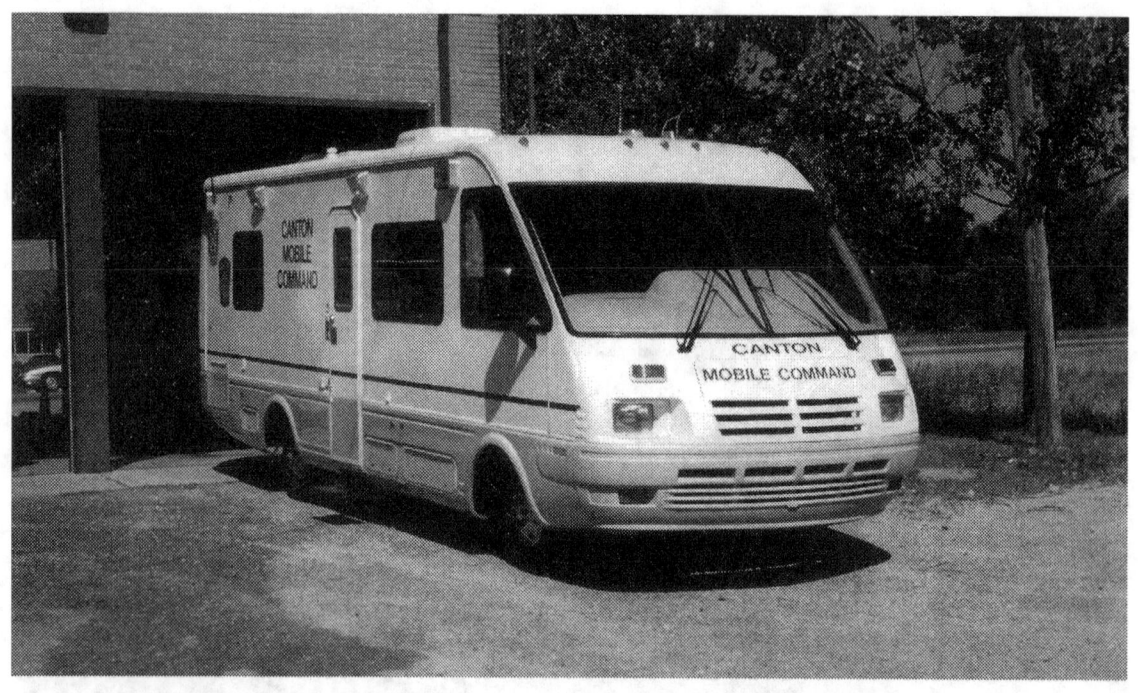

This 1992 National RV on a GM chassis was the second vehicle to be utilized as a mobile comand post by Canton Township. This vehicle was designed to be utilized by both the fire and police departments. Photo by Anthony J. Rzucidlo.

In 2004, Canton Township (Michigan) Department of Public Safety placed this Freightliner / LDV into service. This is the third unit to serve this community as a command unit and to date is the most impressive as it relates to onboard equipment and technology. Sergeant Hardesty (police) who oversaw this project had the unit set-up to be used as an Emergency Operations Center if necessary.
Photo by Anthony J. Rzucidlo.

Carbondale Fire Department (Colorado) assigned this four door 2004 Ford pick-up with an enclosed box area to the duty chief to be utilized as an incident command unit as noted by the lettering on the front fender. Photo by Dennis Maag.

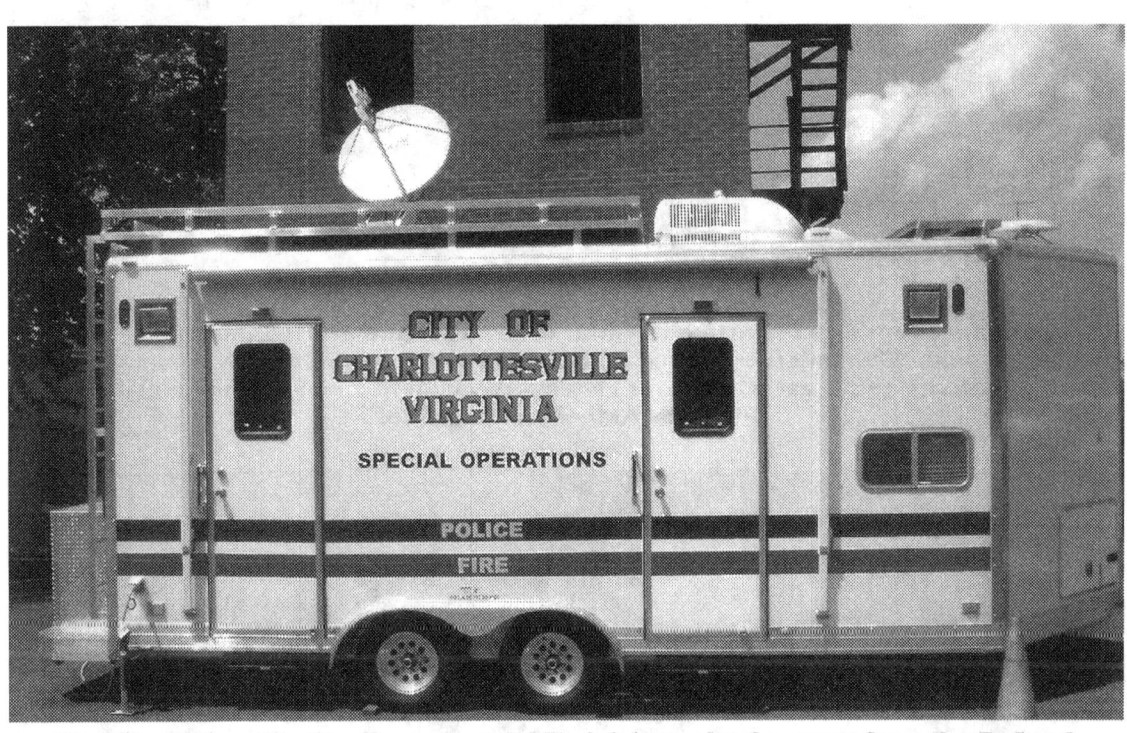

The Charlottesville Fire Department (Virginia) received a grant from the Federal Emergency Managment Agency to purchase this "Special Operations" trailer which can be utilized by either the police / fire departments as a mobile command center and communications unit. This unit responded to the Gulf Coast shortly after Hurricane Katrina struck the region. Photo courtesy Charlottesville Fire Department.

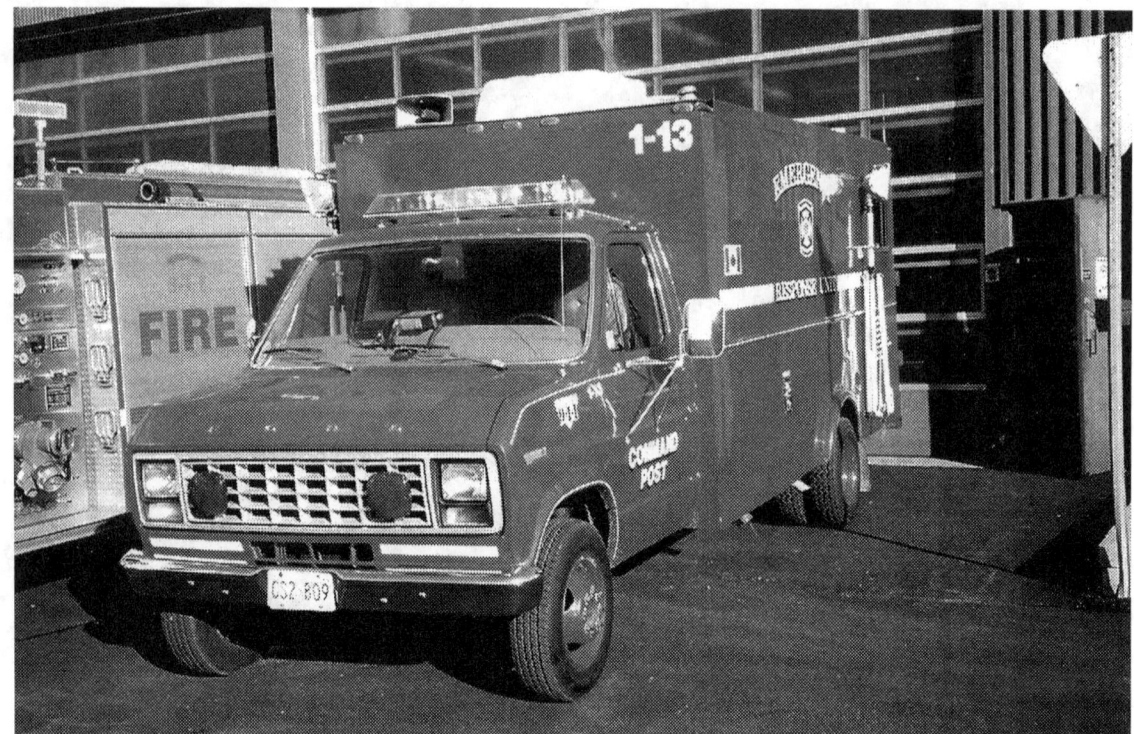

Chatham / Kent Fire Department (Ontario, Canada) still utilizes this 1980 Ford command post. This unit is equipped with the normal equipment for a command post and features a full rear window for viewing of incident scenes when possible.
Photo by Anthony J. Rzucidlo.

This late model Dodge Ram pick-up is used by the Chatham / Kent (Ontarion, Canada) Police Department to pull it's command post, which happens to be a trailer. Not only is the pick-up equipped with emergency lights, but the trailer as well.
Photo by Anthony J. Rzucidlo.

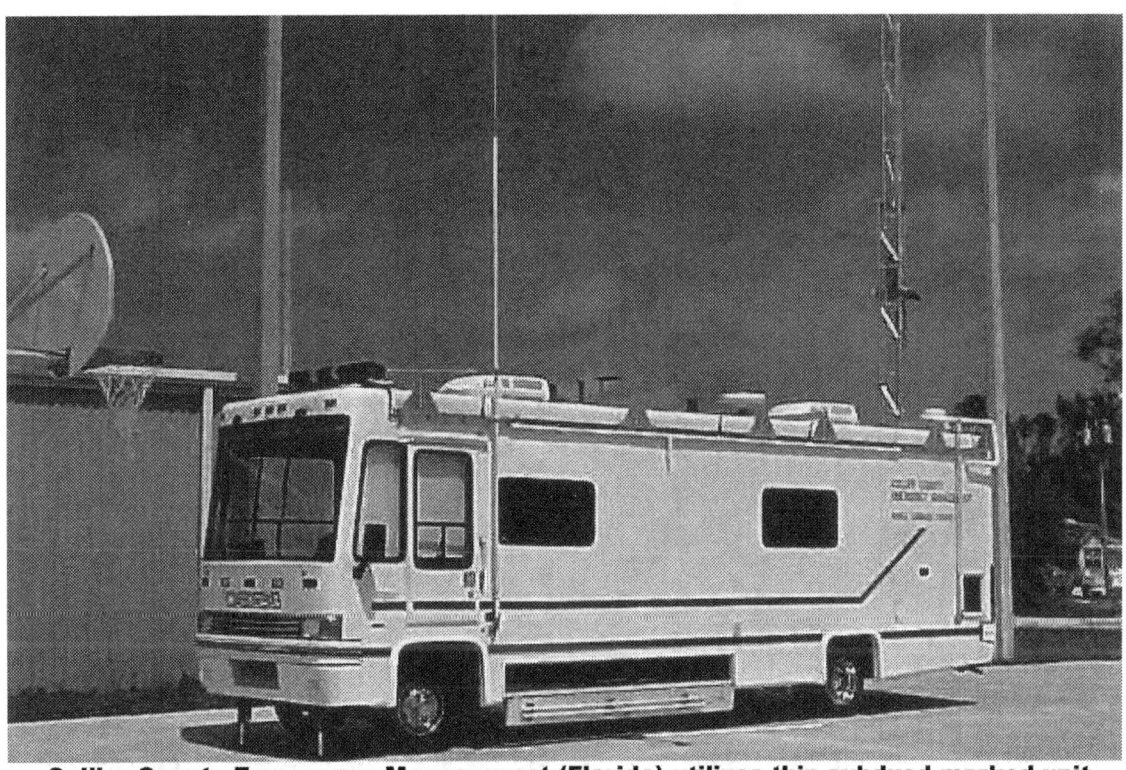

Collier County Emergency Management (Florida) utilizes this subdued marked unit as a "Multi Agency Command, Control & Communications" vehicle. On the front of this unit in reverse lettering is the word "Emergency" in red letters. Photo courtesy Collier County Emergency Managment.

Copper Mountain Ski Resort located in Copper Mountain (Colorado) has operated an in-house fire department since the 1970's. This 1997 AMC / General Hummer / Sutphen which was delivered to the resort in 1998 is not only capable of being a pumper, but it also serves as a command post. Photo by Dennis Maag .

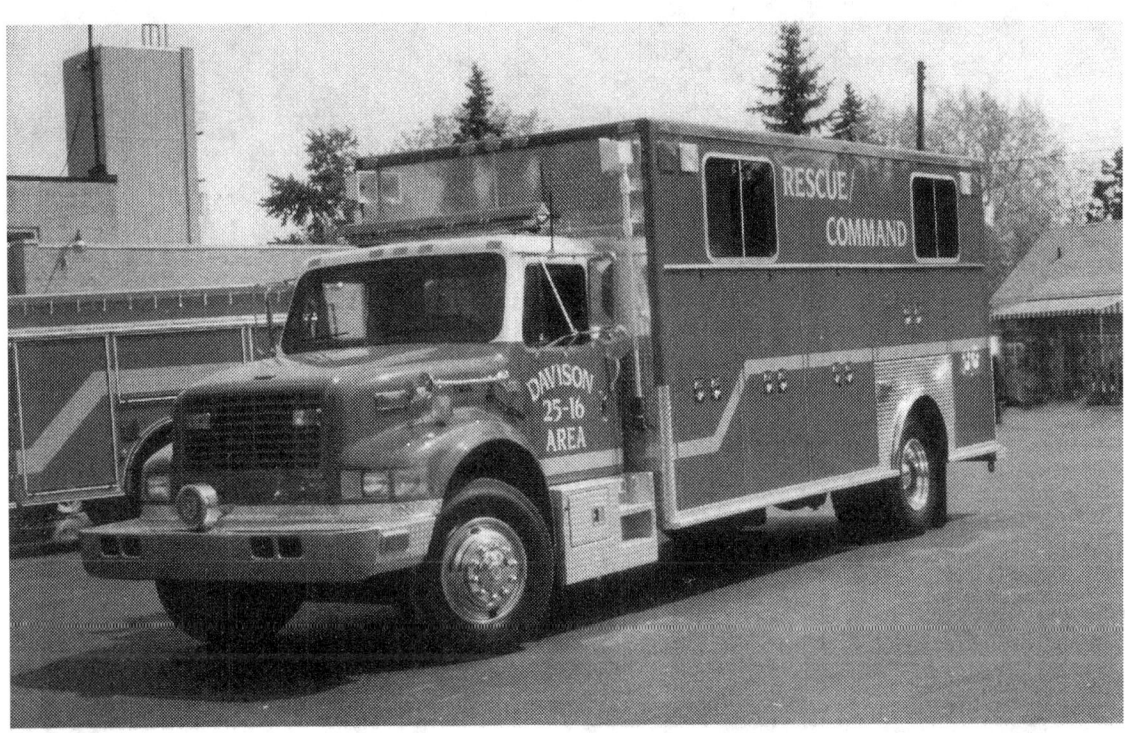

Davision Area Fire Department (Michigan) in 1996 placed this International / American combination rescue / command post into service. Primarily a rescue unit, space was allotted for a command area. Photo by Anthony J. Rzucidlo.

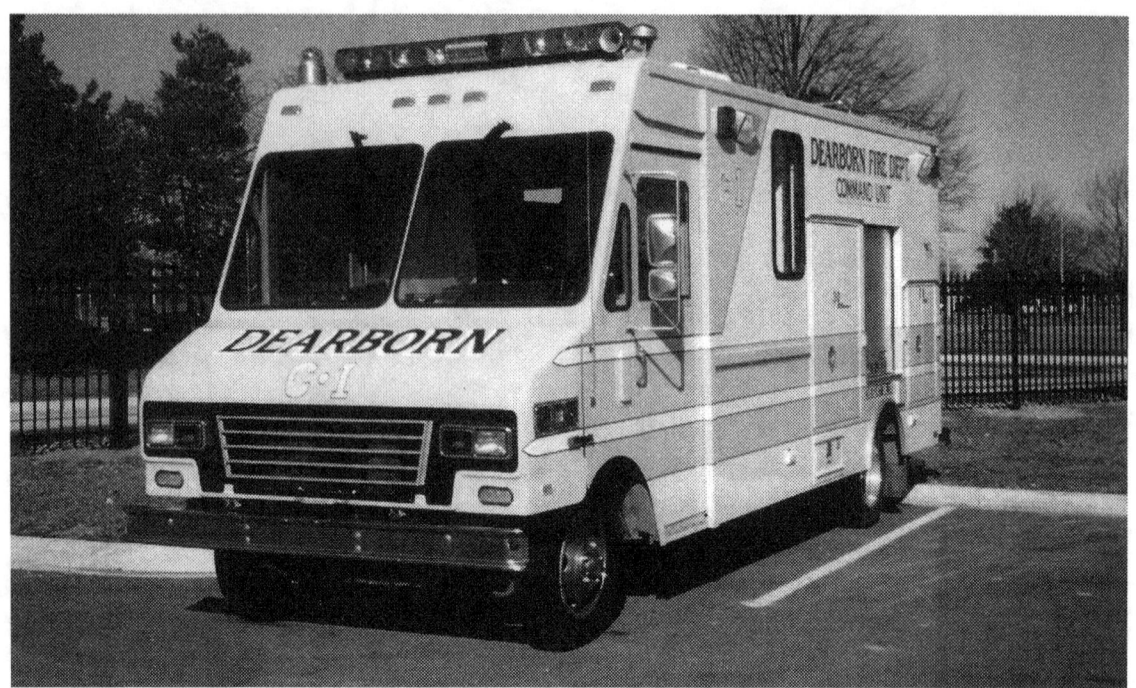

Dearborn Fire Department (Michigan) designed this unit to be utilized as a hazardous materials truck. However, over the years this 1988 Chevy / Utilmaster was converted into a command unit for the department. This unit was taken out of service in 2005. Photo by Anthony J. Rzucidlo.

When responding to initial calls, "Battalion Chief's" of the Dearborn Fire Department respond in this 2003 Ford Expedition. A green incident command light is affixed to the top of the light-bar so, when activated the location of the command post and Incident Commander will be known. Photo by Anthony J. Rzucidlo.

This trailer was donated to the Dearborn Police by the Exchange Club of Dearborn, Michigan. The department utilized the trailer for various events throughout the city as a command unit. This trailer was taken out of service in 2000 when a new command post was purchased. Photo by Anthony J. Rzucidlo.

The Dearborn Police Department's current mobile command post is a 2000 Matthews trailer. The paint scheme and markings on the trailer match that of the departments patrol car fleet. Since the fire department did away with it's command post, they also can utilize this unit if required for an incident. Photo by Anthony J. Rzucidlo.

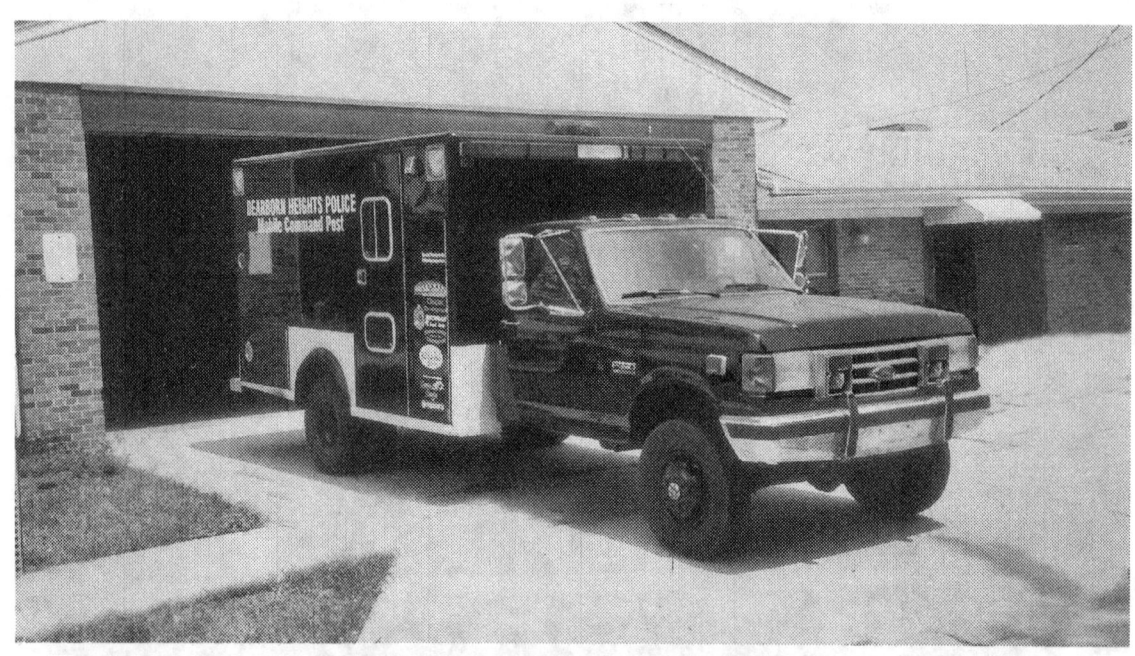

Dearborn Heights Police (Michigan) obtained this former U.S. Air Force ambulance as a surplus vehicle and had it converted into a command unit. The departments SWAT unit primarily utilizes the vehicle. This 1989 Ford was painted black from it's original white color and the local businesses that assisted in the conversion of the vehicle are noted on the side of the unit. This unit is still in service in 2006. Photo by Anthony J. Rzucidlo.

"Battalion Chief's" in the Detroit Fire Department (Michigan) respond to calls and utilize vehicles like this 2002 Chevrolet Tahoe as a command unit. The department does have a command unit that if required can be called out for extended operations. Photo by Anthony J. Rzucidlo.

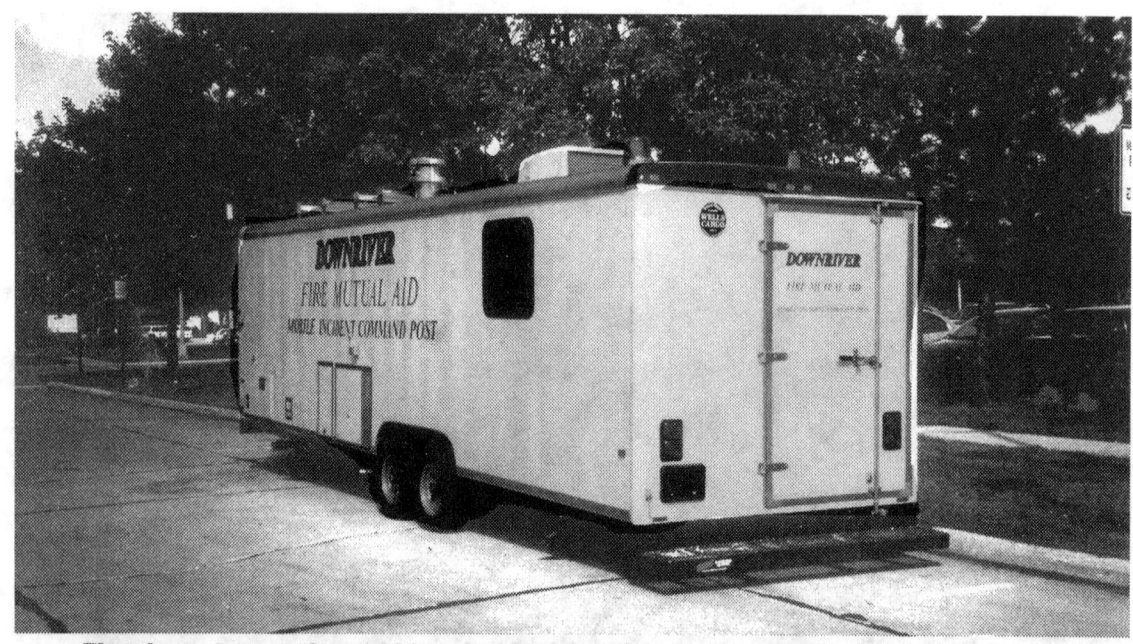

Fire departments located in what is refered to as "Downriver" Wayne County, Michigan (Metro Detroit) purchased this 1997 Wells Cargo trailer and had it made into a mobile incident command post. Photo by Anthony J. Rzucidlo.

East Tawas Fire Department (Michigan) utilized this four door cab GMC chassis as a rescue / command unit. This can be noted by the placement of the green incident command light on top of the light-bar. Photo by Anthony J. Rzucidlo.

Region V, U.S. Environmental Protection Agency Emergency Response Team, responds to environmental incidents with this Ford chassis mobile command post. When this photo was taken, it was staged in Dearborn, Michigan.
Photo by Anthony J. Rzucidlo.

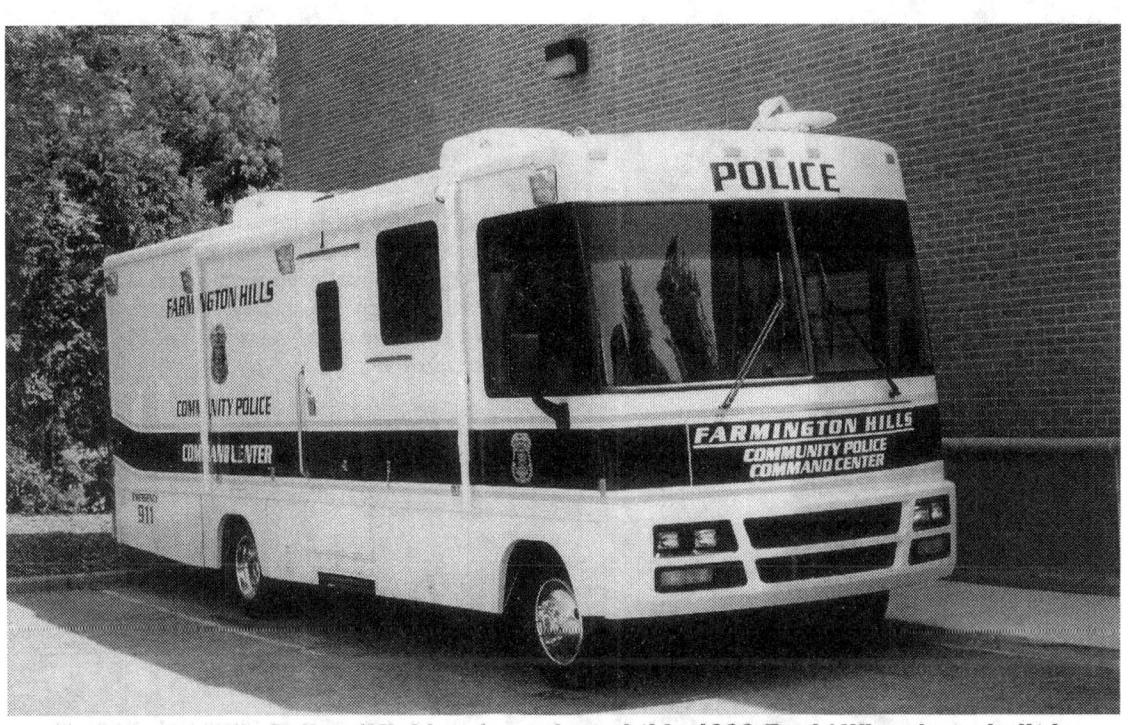

Farmington Hills Police (Michigan) purchased this 1999 Ford / Winnebago built by Faber Specialty Vehicles by utilizing grants and drug forfeiture money. This 30' vehicle is designed to be utilized as a substation / community policing post in residential areas and can be used for major incidents or tactical situations. Photo by Anthony J. Rzucidlo.

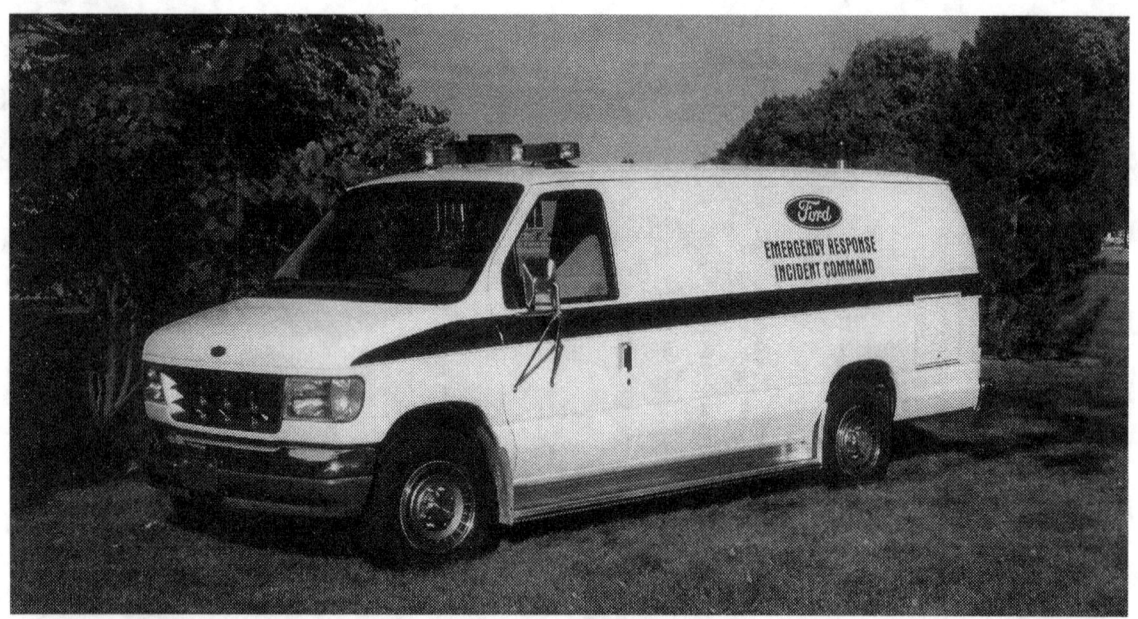

Ford Motor Company Research and Engineering Center in Dearborn (Michigan)
is where this 1994 Ford E-350 saw service. This unit was equiped for confined space
rescue and limited use as a mobile command post. Originally this vehicle was red
and was painted white. The author of this publication was primarily responsible for
this vehicle from 1994-2000. This vehicle was taken out of service in 2002.
Photo by Anthony J. Rzucidlo.

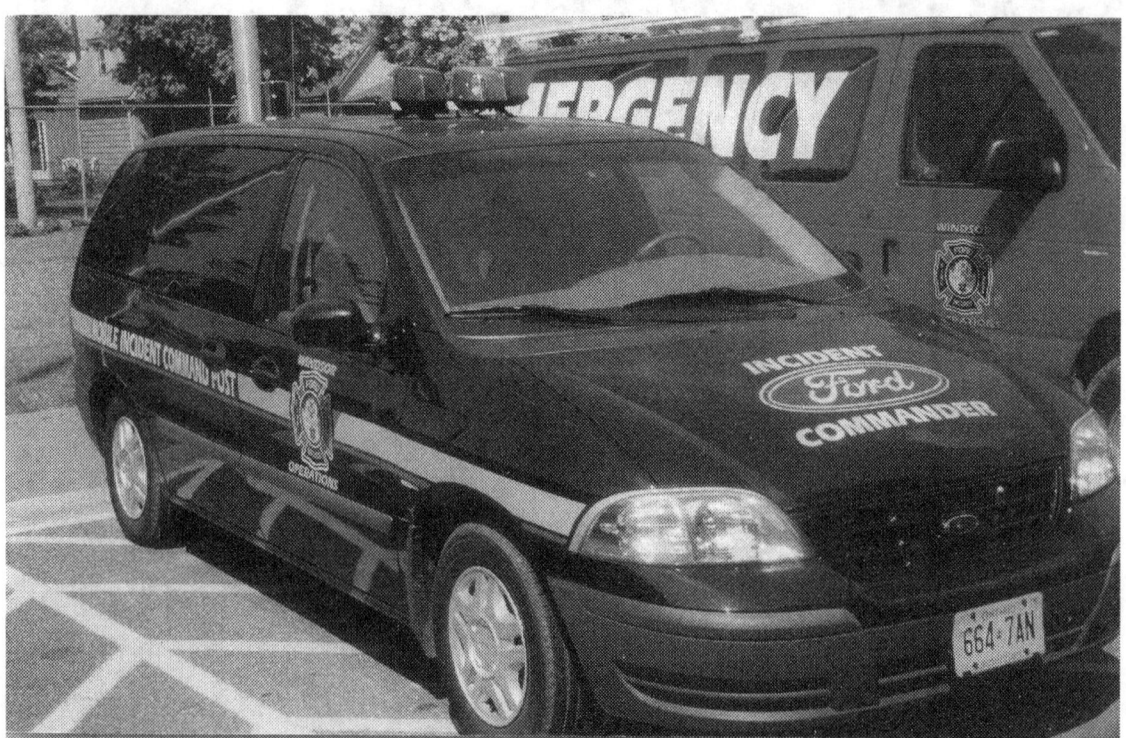

Security Operations at Ford Motor Company facility located in Windsor (Canada)
converted this 2000 Ford Windstar into a mobile command post. Located in the rear
section of the vehicle, a workstation was installed so that the Incident Commander
could track the progress of any incident. Photo by Anthony J. Rzucidlo.

Galesburg Police (Illinois) utilized this Chevrolet step-van as a command post for a variety of events including the popular "Railroad Days." Note the older style type of warning equipment and the placement of spotlights on the roof. This unit was painted in a nice two tone color of a light blue roof and dark blue body. Photo by Anthony J. Rzucidlo.

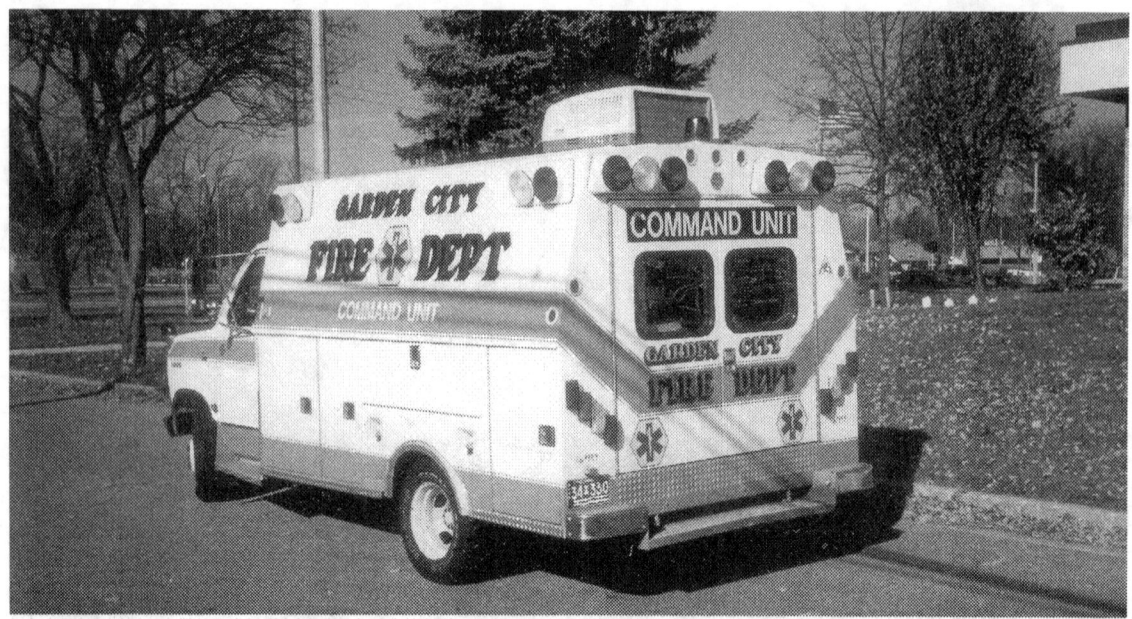

Here is a photo of the Garden City (Michigan) Fire Department command unit while still in service. Some of the features of this unit include the placement of the roof-mounted air conditioner, command light and the lettering on the door identifying the unit as a "Command Unit." Photo by Anthony J. Rzucidlo.

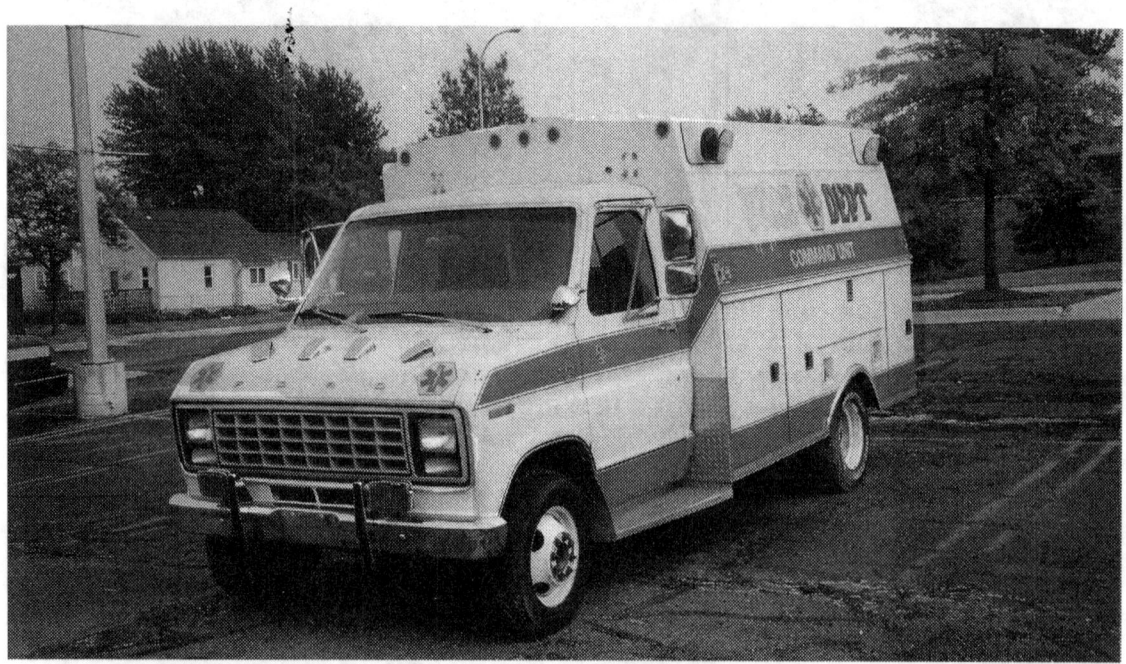

This Ford / Brunn served the Garden City (Michigan) Fire Department as a rescue unit. However, in later years it was turned into a command unit for the fire department. Due to the lack of manpower to bring this unit to various scenes in 2005 it was taken out of service.
Photo by Anthony J. Rzucidlo.

General Services Administration, Federal Protection Service Police operated this 1996 GMC chassis as a mobile command unit. As is typical with most command post units, this one also has an awning located on the side for additional workspace area. Photo by Christopher M. Taylor.

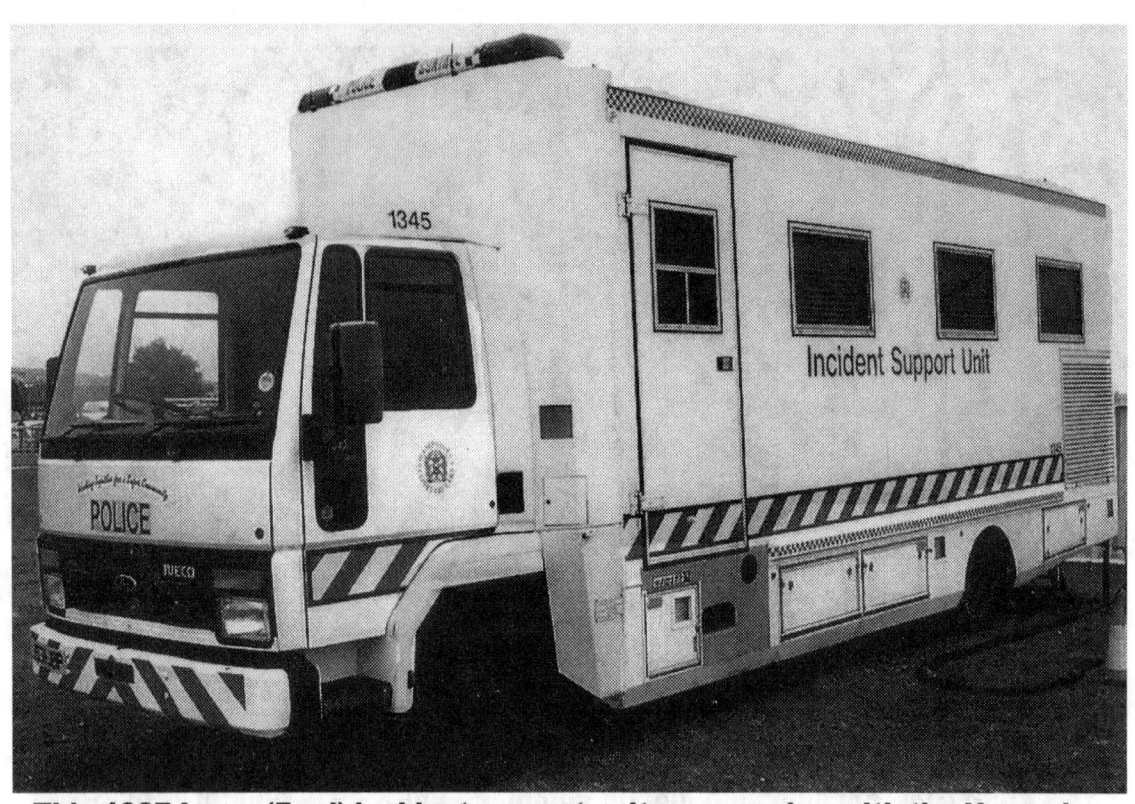

This 1987 Iveco (Ford) incident support unit saw service with the Hampsher Police (United Kingdom of Great Britain). Note the placement of the light-bar on top of the roof of the support unit. Photo by Christopher M. Taylor.

Hoechst-Celanese is a chemical manufacturing company that operates several large industrial facilities in Texas. The facility in Bishop operated this 1983 Ford that was converted by the fire department members into a command post. The unit was white with a blue stripe with the word "Command" in white letters on the rear. Photo by Dennis Maag.

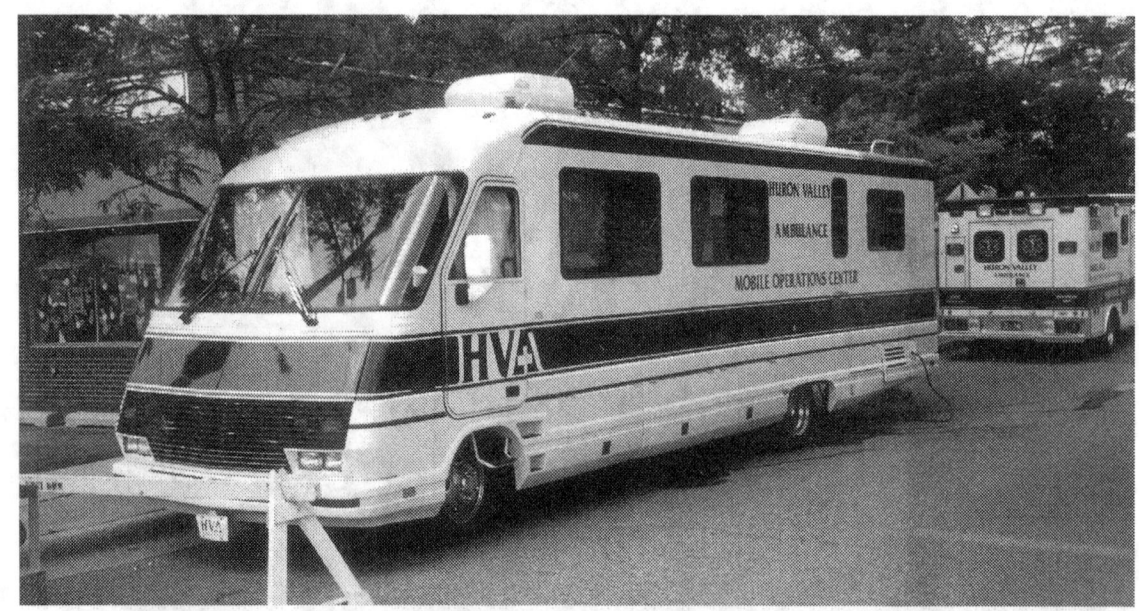

Huron Valley Ambulance (Michigan) operates this 1988 Pace Arrow as a mobile medical operations center for special events including University of Michigan home football games. Plans are being made to replace this vehicle. Photo by Anthony J. Rzucidlo.

The Inkster Police Department (Michigan) obtained this commuter type bus from a local transit authority. The department converted this Ford / Champion into a multi task purpose unit, that being a crime scene and command post. Photo by Anthony J. Rzucidlo.

In 2002 Tompkins County Area Transportation (New York) donated this bus to the Ithaca Police Department who converted the bus into a mobile command post for the Special Weapons & Tactics (SWAT) team. Photo courtesy Ithaca Police Department.

Jackson Police (Michigan) appear to have had this box truck mounted on a Chevy chassis made into a mobile command unit. Note the snow on the windshield of the truck, one of the hazards of not having covered parking for special unit vehicles in winter months.
Photo by Anthony J. Rzucidlo.

It appears as this 1974 Cheverlot step-van has been re-cycled. It appears as this unit started out as a fire department vehicle and the Knoxville Police Department (Illinois) recieved it and made it into a command unit. Photo by Anthony J. Rzucidlo.

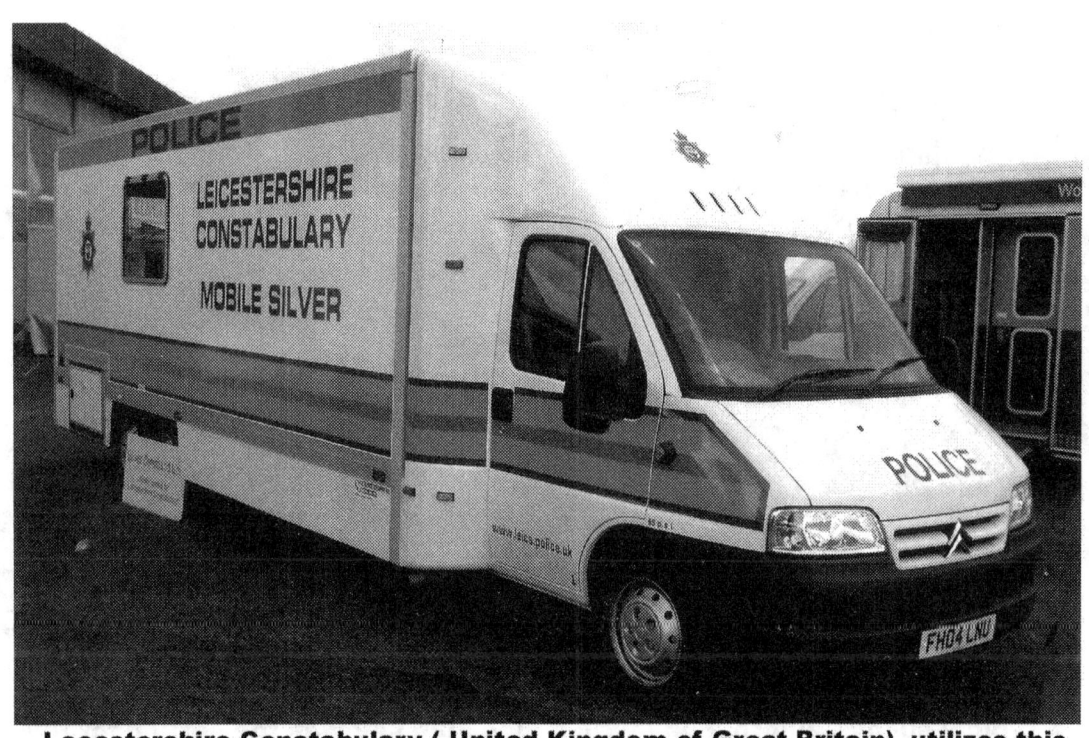

Lecestershire Constabulary (United Kingdom of Great Britain) utilizes this 2004 Peugeota as a mobile command unit. In Great Britain they utilize different terminology as it relates to the command staff, thus "Mobile Silver" indicates where the Incident Commander can be located.
Photo by Christopher M. Taylor.

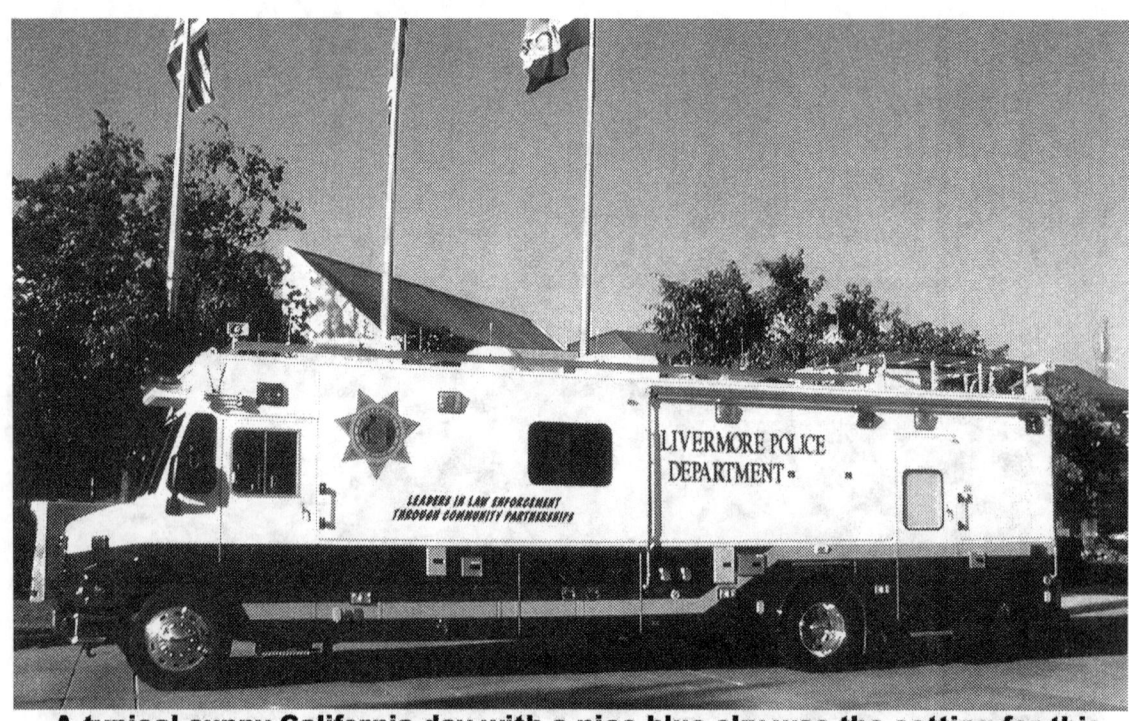

A typical sunny California day with a nice blue sky was the setting for this photo of a LDV command center built for the Livermoore Police Department. This unit was placed into service in 2002, and has a nice paint scheme that consists of white and blue with a gold accent stripe. Photo courtesy Livermore Police Department.

Livingston County Sheriff (Michigan) was responsible for this 1999 GMC step-van that was donated by General Motors and turned into a mobile command unit at a cost of $40,000.00 to the county. In the winter months, arrangments are made for this unit to be housed at the local fire station. Photo by Anthony J. Rzucidlo.

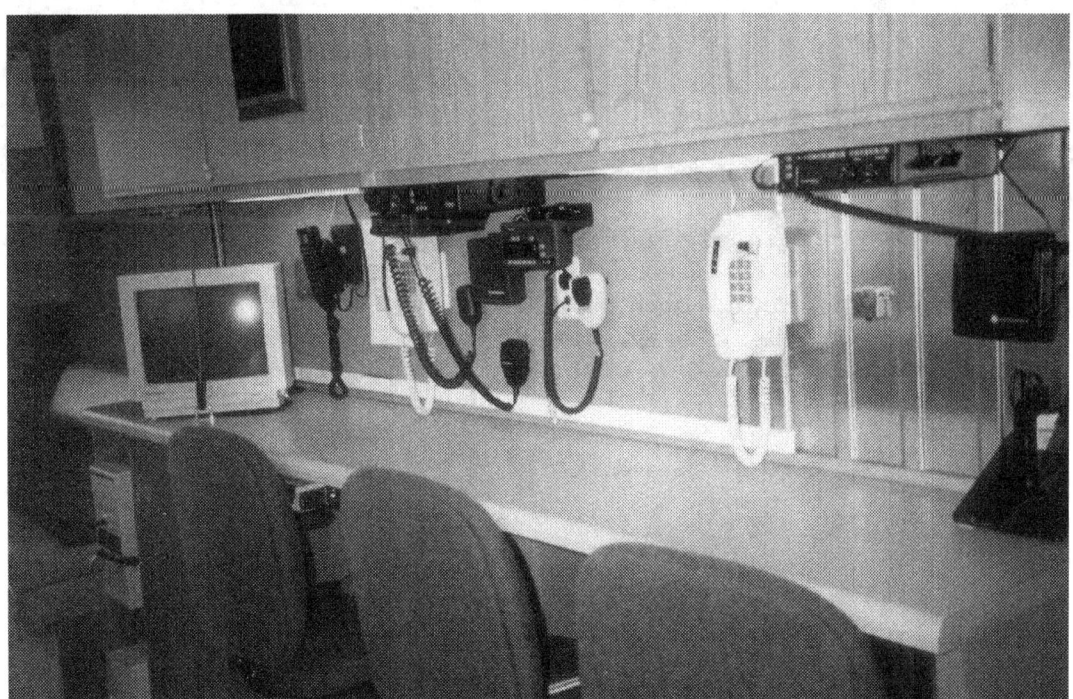

The three-person communications area of the Livingston County Shreiff's command post is shown. The entire area of the interior of this vehicle was planned out well and made functional which included a conference area. Photo by Anthony J. Rzucidlo.

London Police (Ontario, Canada) had the use of this 1993 Foran motor-home as a command center. Interesting thing to note about this vehicle is the viewing platform on the rear of the roof. Photo by Anthony J. Rzucidlo.

London Fire Brigade (United Kingdom of Great Britain) 1985 Range Rover utilized as a major incident vehicle. This unit is equiped with two roof mounted spotlights, somewhat unusual for emergencey vehicles in Britain. Photo by Christopher M. Taylor.

This 1997 Volvo bus chassis with Spectra bodywork is a major incident command and control unit for the London Fire Brigade (United Kingdom of Great Britian). The unit red in color, has a white checkerboard design around the entire vehicle to increase its visibility. Photo by Christopher M. Taylor.

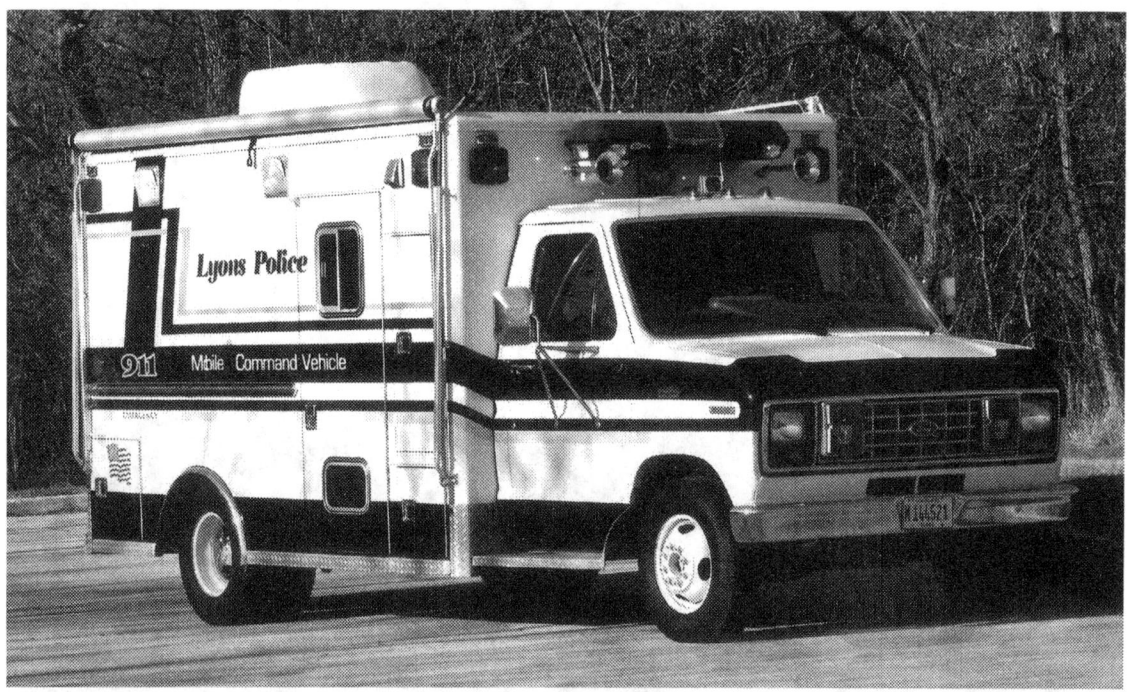

This 1986 Ford / Wheeled Coach started its service career as an ambulance. In 2003 the Lyons Police Department (Illinois) converted the former ambulance into a mobile command vehicle. This unit sports at least five radio antennas on the roof of the command area. This truck is still in service in 2006. Photo courtesy Lyons Police Department.

This rear shot of the Lyons Police Department command vehicle, shows where the Incident Commander and others would gather to manage any incident requiring this unit. Photo courtesy Lyons Police Department.

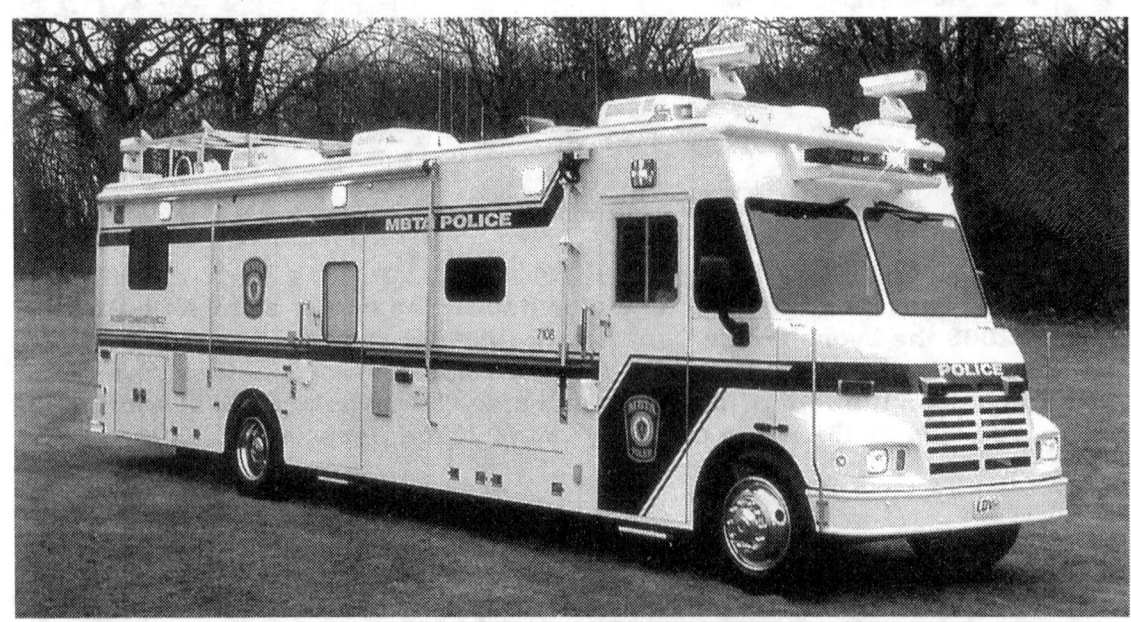

Massachusetts Bay Transit Authority Police (MBTA) utilizes this 1999 International / LDV as an incident command vehicle. This vehicle is somewhat unique that it has cameras mounted on all four corners of the roof. Photo courtesy Massachusetts Bay Transit Authority.

This rather large looking vehicle is a 1994 Leyland with bodywork by Glover was designed to serve as a police conference unit for the Metropolitain Police (United Kingdom of Great Britain). Photo by Christopher M. Taylor.

This major incident vehicle operated by the Metropolitan Police and assigned to Heathrow airport in Great Britain is a 1995 Leyland chassis with body work by Customline. Note how the warning lights are built into the body. Photo by Christopher M. Taylor.

The Metropolitan Police (United Kingdom of Great Britian) placed into service at Heathrow airport this 2004 40' MacNeillie forward support trailer. Sergeant Christopher Taylor, Metropolitan Police, was greatly responsible for the design and development of this unit. Photo by Christopher M. Taylor.

This photo shows the forward support trailer assigned to Heathrow being pulled by a 2000 Iveco (Ford) tractor. Note the yellow warning lights on top of the tractor and trailer. When moving around the airport they are activated so air traffic controllers can see the vehicle moving about.
Photo by Christopher M. Taylor.

Done in the same color as the departments patrol vehicles, the Michigan State Police stationed its 1998 International command post at the Michigan State Fair in Detroit. The Michigan State Fair is the oldest state fair in the United States. Photo by Anthony J. Rzucidlo.

You can't miss what the purpose of this unit is for. This incident command
unit is operated by the Mid & West Wales Fire and Rescue Service
(Great Britain). Note the high visable markings.
Photo by Christopher M. Taylor.

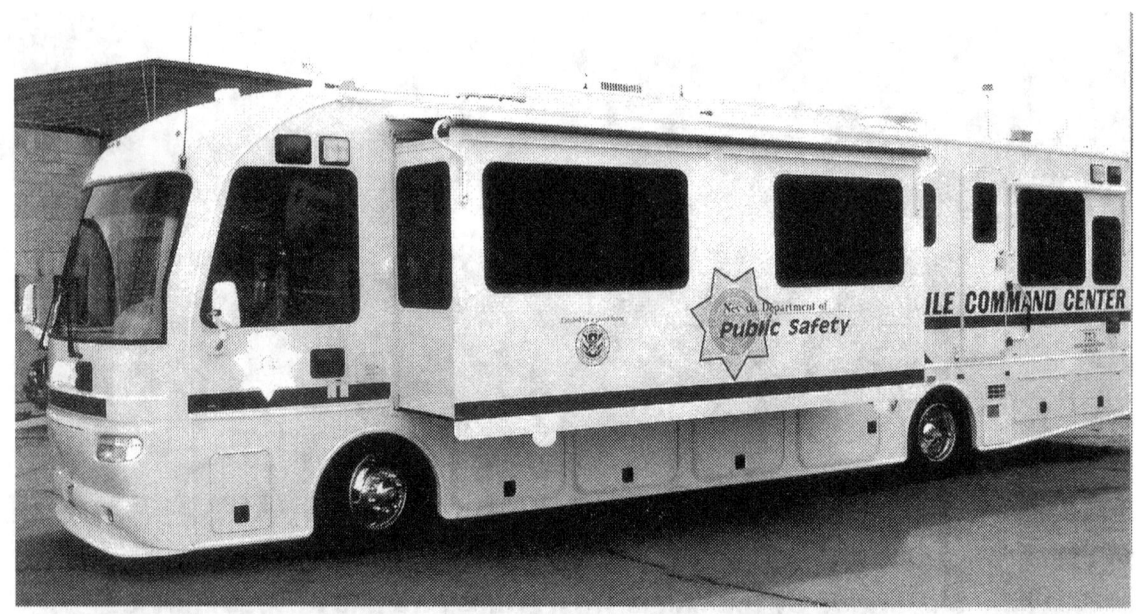

The Nevada Department of Public Safety obtained a grant from Homeland Security for $800,000.00 which they used to purchase two 2005 Alfa See-ya motorhomes. The department had them gutted and converted into mobile command centers. Note how this unit has a side extension that allows for more working space when the vehicle is at the sight of an incident. Photo courtesy Nevada Dept. of Public Safety - Highway Patrol.

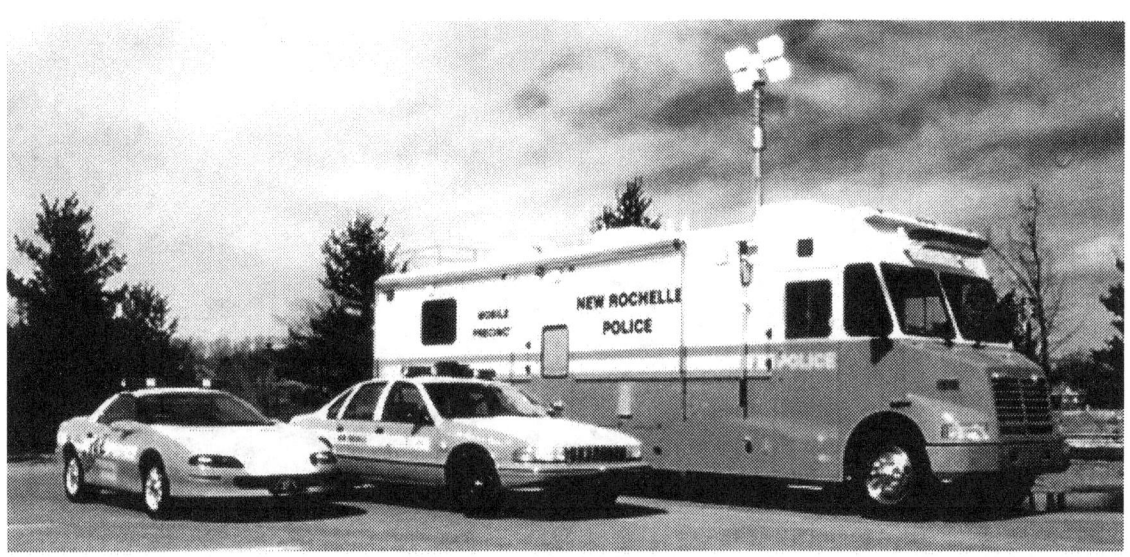

Truck No.1701 belonging to the New Rochelle Police (NewYork) which is 37' long, 7.5' wide and 11' 6" high was obtained through a federal grant. The unit is designated to be used as a mobile precinct command center. The truck is painted in an attractive paint scheme of white and light blue, which is identical to the paint scheme of the departments patrol vehicles. Photo courtesy New Rochelle Police Department.

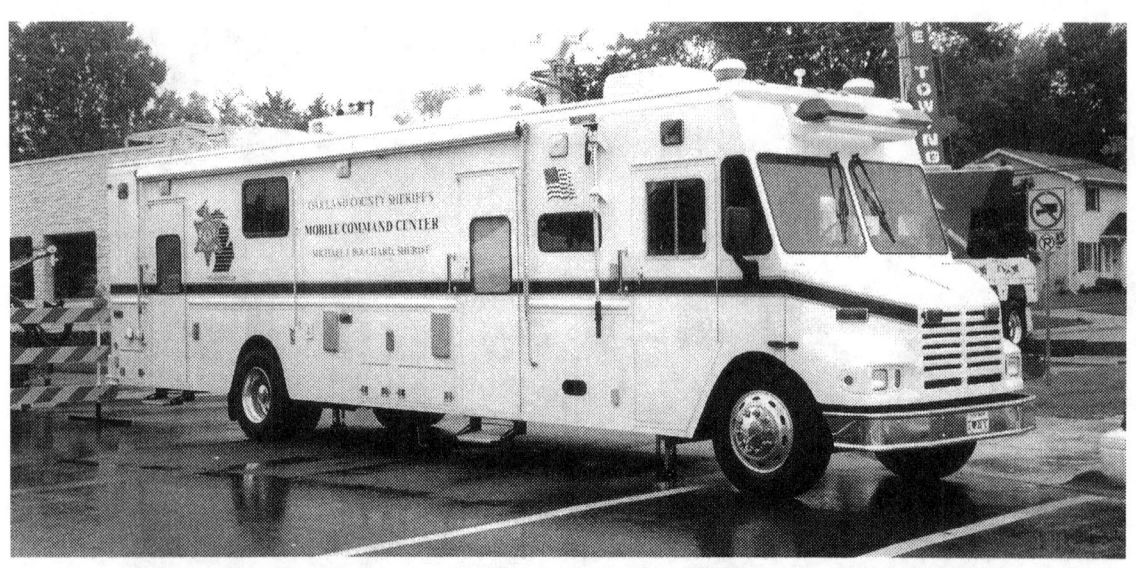

The Oakland County Sheriff Department (Michigan) has this 2002 Kenworth / Grumman in their vehicle fleet, which is utilized as a mobile command center. Note the typical Michigan sheriff department decal near the rear of the unit. This unit is only one of several specialized vehicles that the department has. Photo by Anthony J. Rzucidlo.

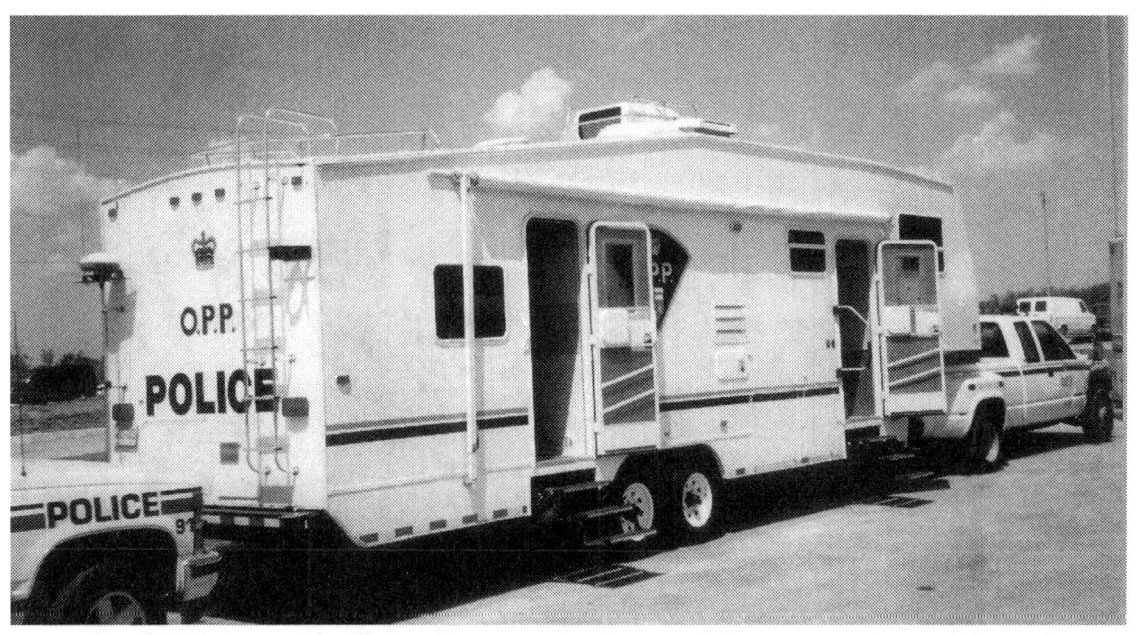

Ontario Provincial Police (Canada) utilizes this trailer for the departments command post. On the other side of this trailer is a pull out section that affords those working inside more room. The trailer is attached to a department pick-up that is used to transport it to various areas.
Photo by Anthony J. Rzucidlo.

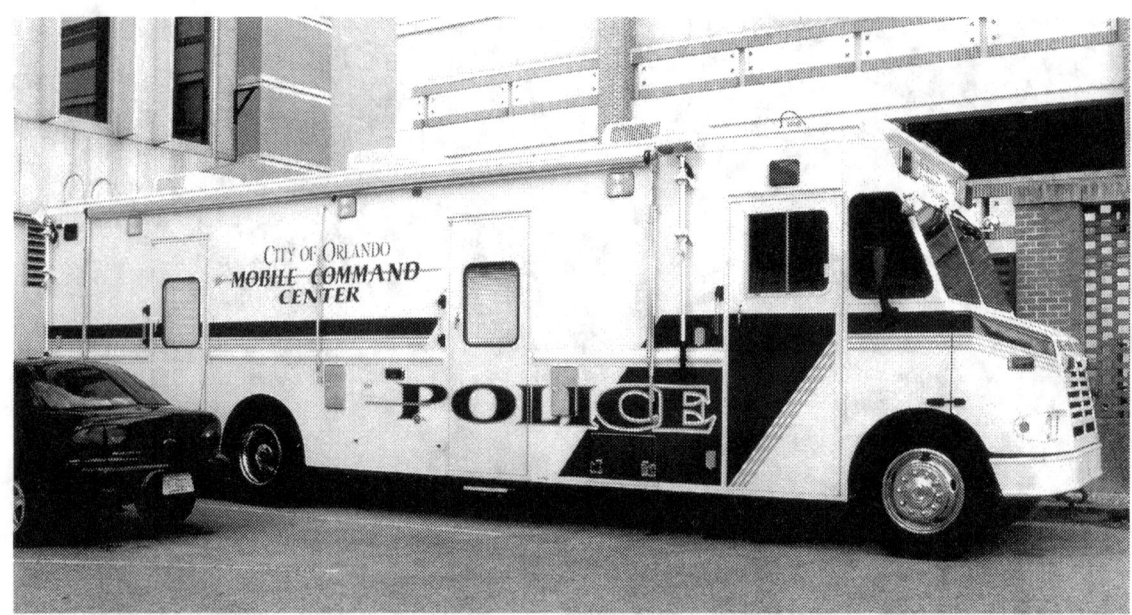

This 1998 International / Grumman command center is operated by the Orlando Police (Florida). The unit is mostly white with dark blue and gold accenting. Interesting feature on this unit is the placement of spotlights on both sides of the cab area. Photo courtesy of Orlando Police Department.

This close-up photo of a 1998 GMC Suburban was utilized by the Port Authority Police Department (New York) at JFK International Airport as a mobile command post. Note the lettering on the rear side window indicating that fact. Photo by Christopher M. Taylor.

With the Rose Bowl stadium in the background and the mobile command center of the Pasadena Police Department (California) in the forground makes for a rather interesting photo. The command unit was manufactured by LDV. Photo courtesy Pasadena Police Department.

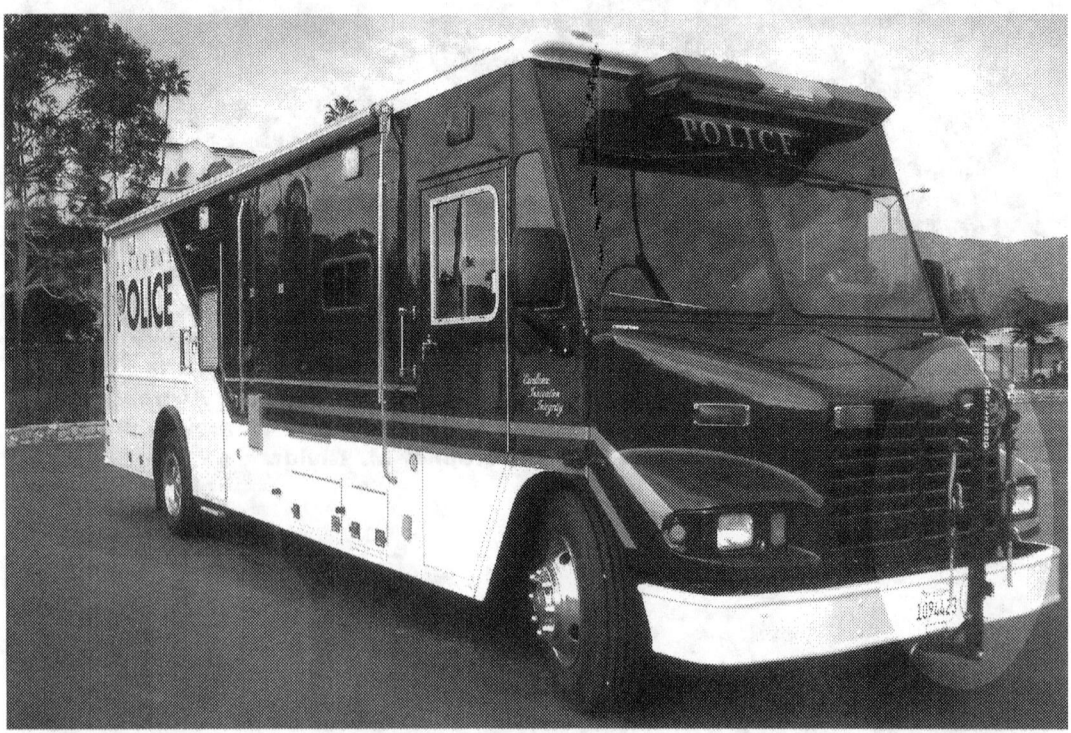

A closer view of this Pasadena command unit reveals that it is painted in an attractive paint scheme of blue and white with a gold accent stripe. Also in the word police on the side of the unit in the center of letter "P" is a red rose. Also this unit is equipped with a bike rack on the front and the rear to transport the departments bike patrol. Photo courtesy Pasadena Police Department.

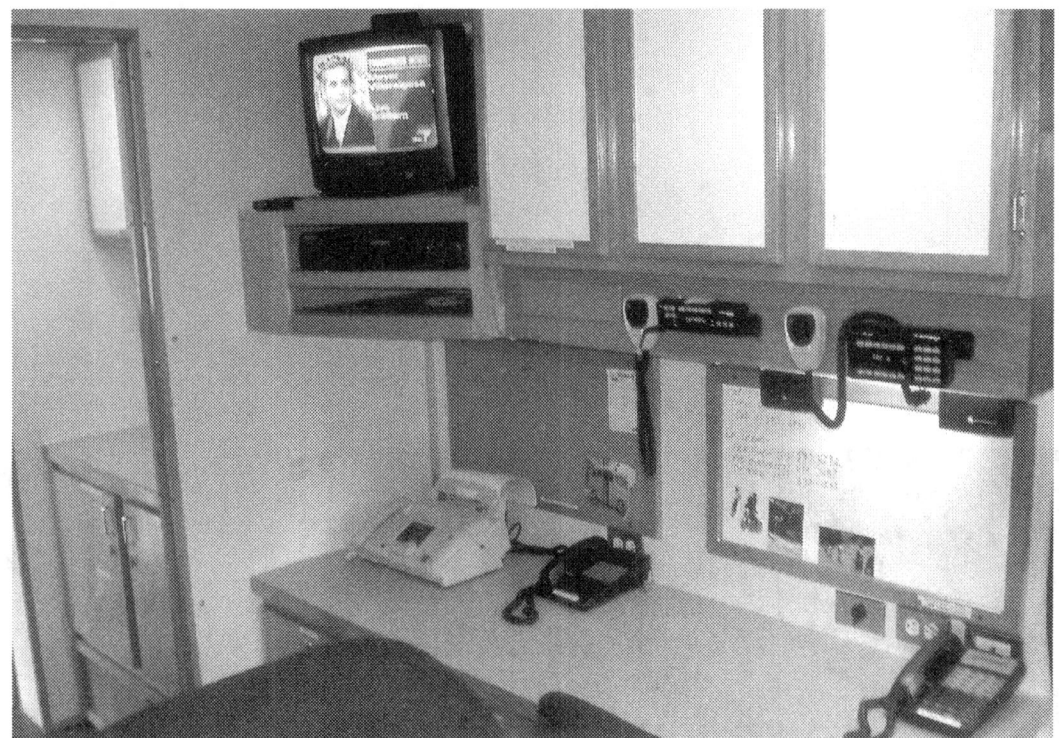

This interior photo shows the communications set-up of the Pasadena unit. In this photo you see a neat and clean set-up for the radios, as well as plenty of whiteboards for note taking. Also this unit is equipped with a fax machine and a television to monitor local news activities.
Photo courtesty of Pasadena Police Department.

Here is a photo of the conference area for the Pasadena command unit. Conference areas are important to have, so that the command staff can have an area to meet during an incident in a somewhat private area.
Photo courtesy Pasadena Police Department.

The " Battalion Chief" on duty at the Redford Township Fire Department (Michigan) utilizes this 1995 GMC Vandura 2500 as a command unit. This unit has a work station set-up in the passenger area of the van. This unit is still in service as of 2006. Photo by Anthony J. Rzucidlo.

Rowan County Telecommunications (North Carolina) obtained this 40' Thomas
bus from a local hospital for one dollar. This command unit is primarily
utilized as a communications center throughout the county. It is equiped
with over 22 diferent strobe and LED lights.
Photo courtesy Rowan County Telecommunications.

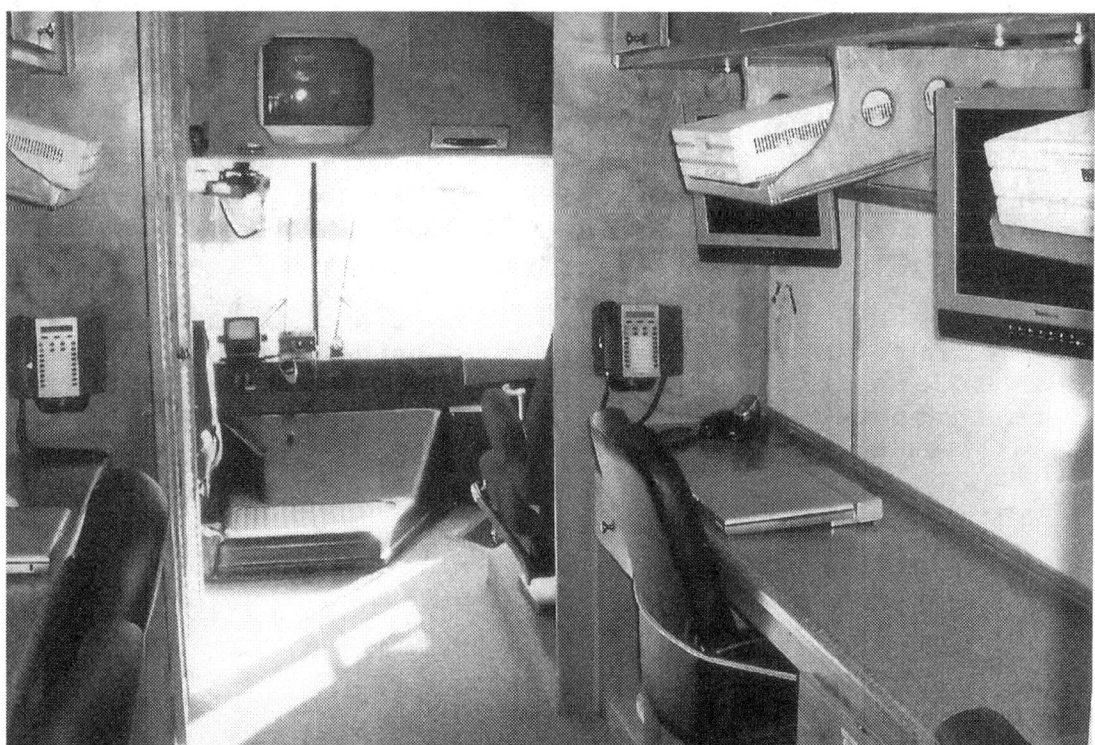

Here is an interior photo of the Rowan County Telecommunications command
unit showing some of the equipment at one of the workstations, plus a view
into the drivers compartment.
Photo courtesy Rowan County Telecommunications.

Royal Air Force (United Kingdom of Great Britain) had this 1991 Land Rover assigned to them and was utilized as a mobile operations vehicle. Note the interesting placement of the spare tire on the hood and the radio antenna on the left front fender. Photo by Christopher M. Taylor.

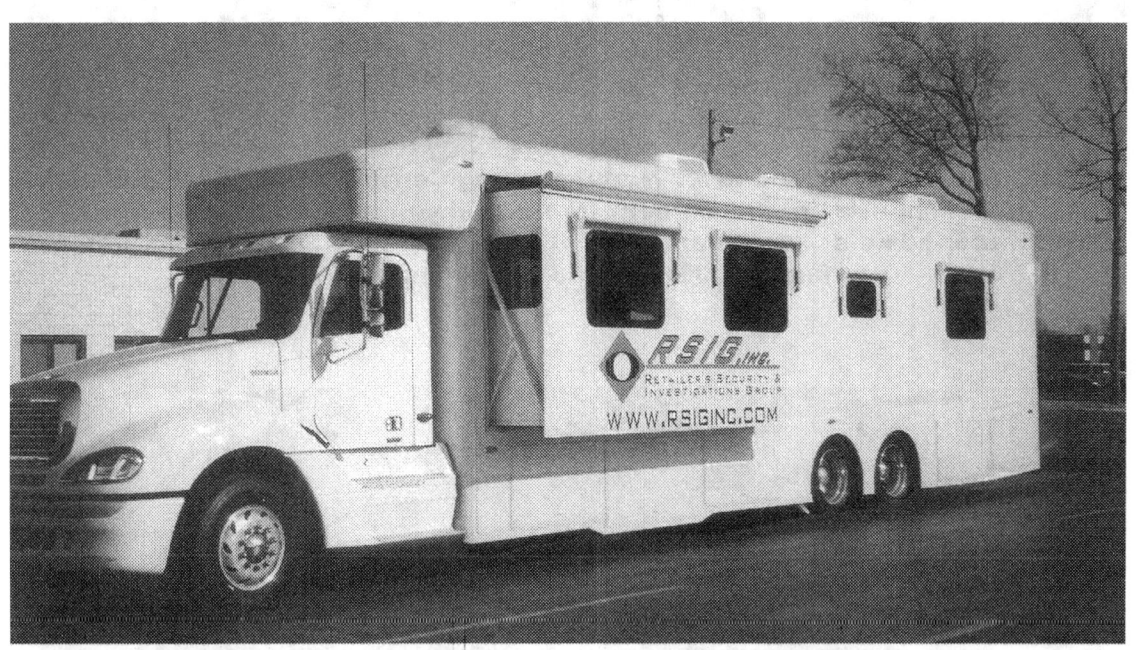

RSIG Security Group, Inc., based out of Southfield (Michigan) utilizes this 40' 2004 Freightliner as a command post for the many special events that they provide coverage for. This is another unit that has a slide out area for additional workspace. Photo by Anthony J. Rzucidlo.

Shell Oil Company Wood River Manufacturing Center - Wood River (Illinois) has had an in-house fire department for several years, and operates this 1977 GMC motor-home as "Command Post 1" This vehicle was still in service as late as 2004. Photo by Dennis Magg.

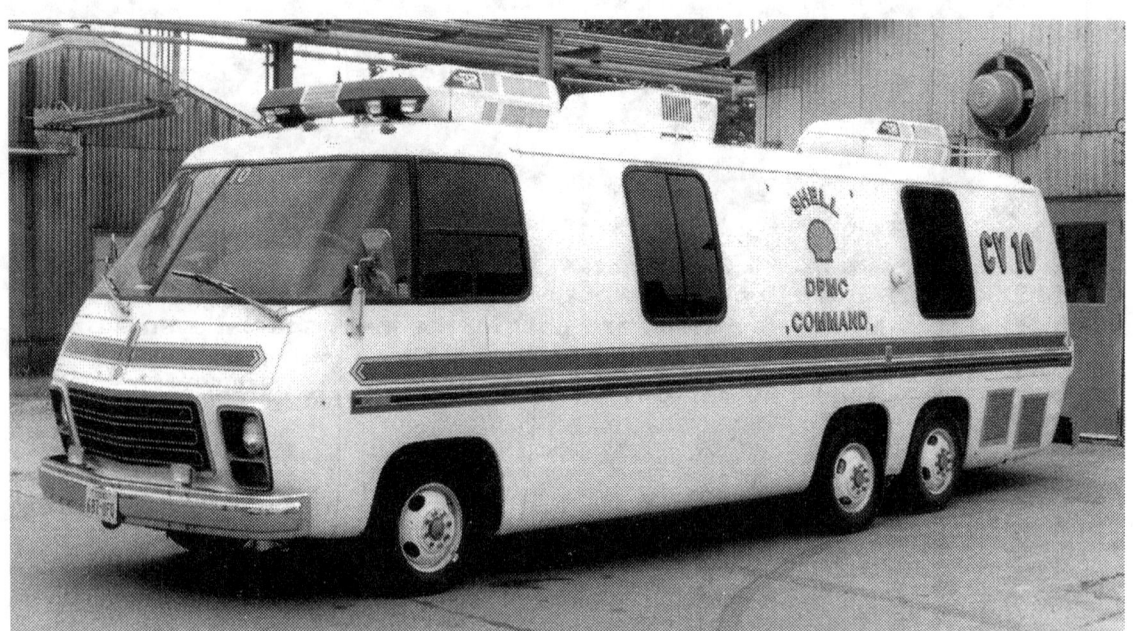

Shell Oil Company Deer Park Manufacturing Center--Deer Park (Texas) operated this 1978 GMC motor-home as a command post, until the mid 1990's when it was replaced with a Blue Bird manufactured unit.
Photo by Dennis Maag.

St. Louis Fire Department (Missouri) placed this 2004 Freightliner/Utilimaster into service as a mobile command post. Besides scene lights built into the unit, it also has a portable floodlight unit affixed to the side of it for additional lighting at the scene of incidents. Photo by Dennis Maag.

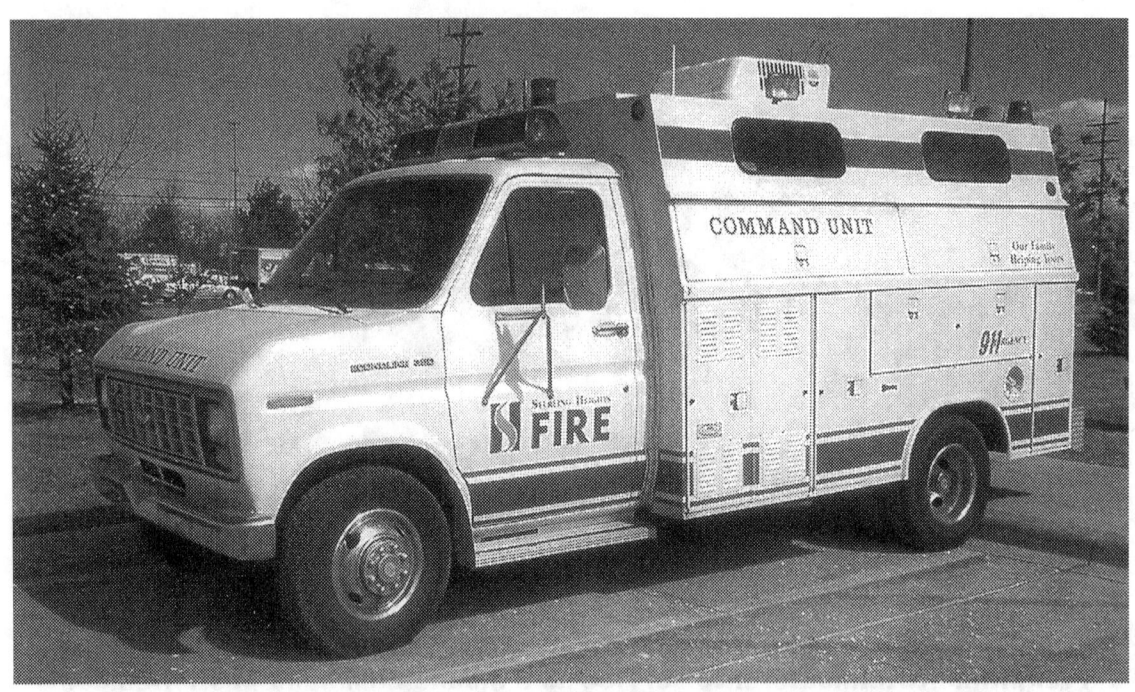

Sterling Heights Fire Department (Michigan) operated this Ford as a command unit. It appears as this unit was used origianlly as an ambulance or a rescue unit. Recently, this unit has been placed in reserve status and is used for other assignments. Photo by Sherman Rice.

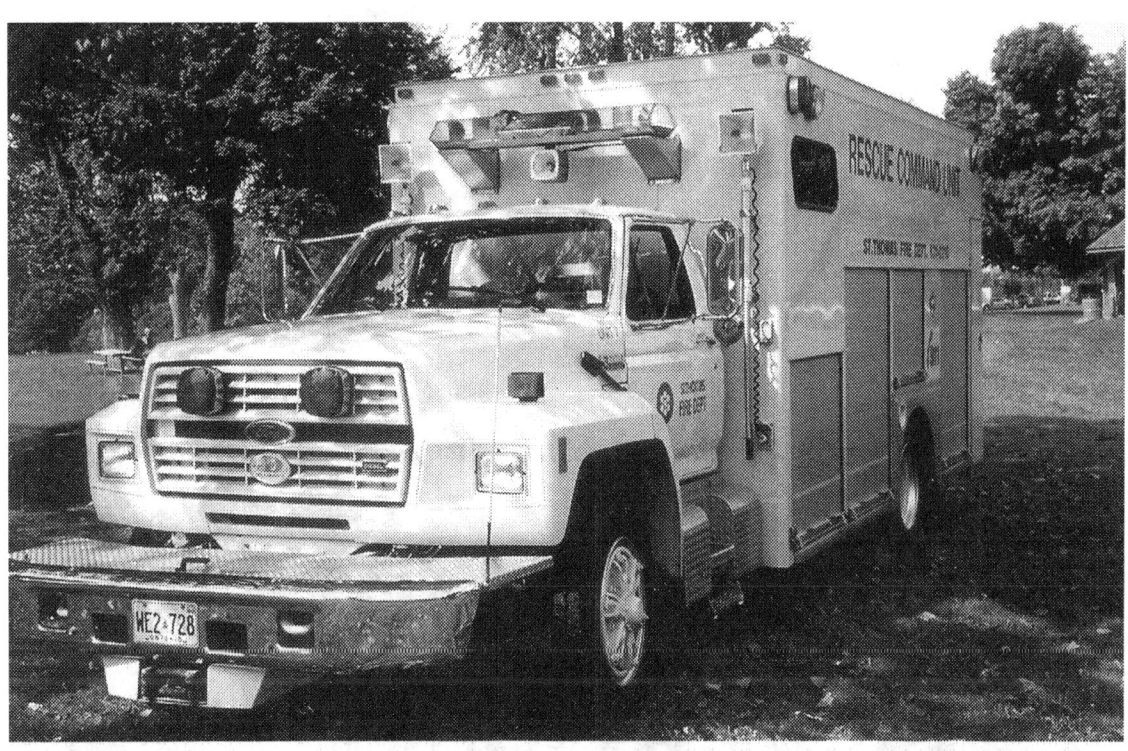

Photographed at a fire apparatus muster in 2004 is this Ford/F800 chassis combination rescue/command unit of the St. Thomas Fire Department (Ontario,Canada). Some interesting features of this unit is the wench mounted underneath the front bumper and the portable flood lights mounted on the sides of the box area. Photo by Anthony J. Rzucidlo.

This 1967 Bedford RL 4x4 with the body work done by Marshall's of Cambridge was utilized by the Sussex Police (United Kingdom of Great Britian) at Gatwick airport as a major incident vehicle. Note the mud on the tires, apparently this vehicle did see some off-road usage. Photo by Christopher M. Taylor.

This unit is a 1997 Mercedes Benz 4x4 with chassis by Kinetic Engineering Bodywork is utilized by the Sussex Police (Great Britain) at Gatwick airport as a major incident vehicle. This vehicle is painted a nice shade of blue with red, yellow and checkerboard striping down the side of this unit.
Photo by Christopher M. Taylor.

Swanton Fire Department (Ohio) has this International/Michigan First
Response vehicle and utilizes it as a combination rescue/command unit. On
the roof of the vehicle are two strobe lights, one green for command post
identification and the other blue for use when directing helicopters that are
utilized to transport injured persons involved in vehicle accidents.
Photo by Anthony J. Rzucidlo.

In addition to the above-mentioned vehicle, the Swanton Fire Department
located in the northwest part of Ohio has this Chevy Tahoe that responds as
the primary command unit. Across the front of the windshield appears the
lettering that reads "Swanton Command ". Photo by Anthony J. Rzucidlo.

This 4x4 is a 1989 Ford Transit rather plain looking compared to other units, assigned to the Thames Valley Police (United Kingdom of Great Britain) and was used by the department's headquarters operations as a command vehicle. Photo by Christopher M. Taylor.

Another type of unique unit that has been utilized by the Thames Valley Police was this major incident trailer. This is a very interesting and compact unit. On the front of the trailer a notice is posted indicating that this unit can only be towed by Land Rovers. Photo by Christopher M. Taylor.

Union Carbide Corporation - Texas City (Texas) utilized this 1989 Ford van as a command post. In fact when this photo was taken in 1993, the "Emergency Director" of the in house fire dept., was assigned to the vehicle. Note how the green command post light has been mounted high above the vehicles light-bar. Photo by Dennis Maag.

Universal Ambulance (Michigan) has this 1974 GMC chassis special events unit. The vehicle has an interesting paint scheme of white over orange. Also it's interesting to note how the side awning has been installed. This unit was still around in 2005. Photo by Sherman Rice.

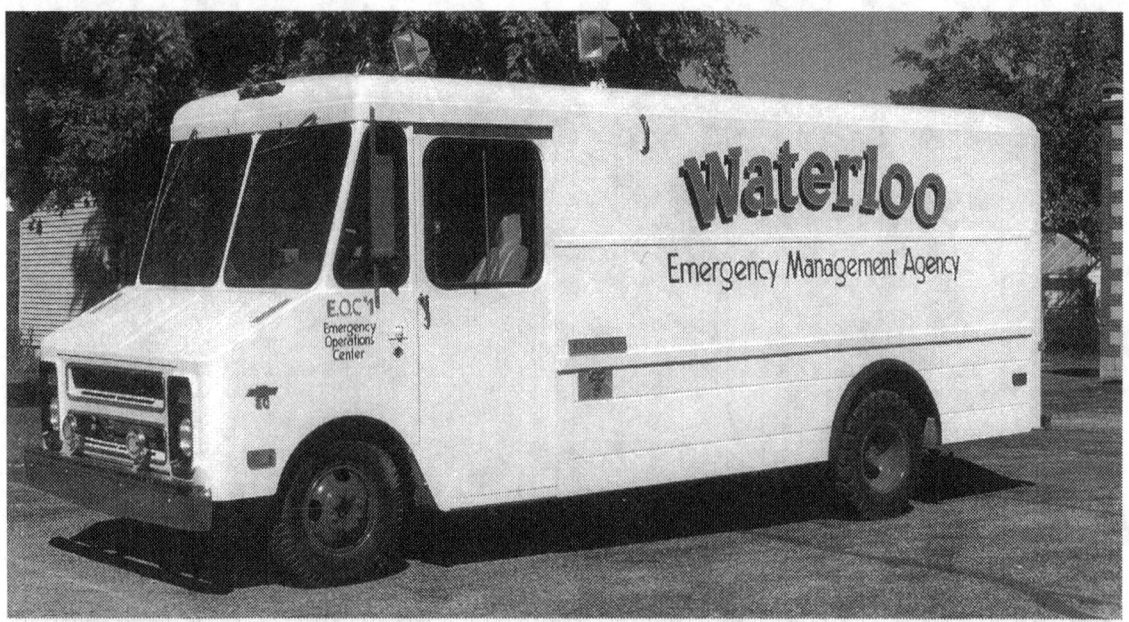

This Chevrolet step-van was converted into an emergency operations center for the Waterloo Emergency Managment Agency (Illinois). Interesting feature of this unit, is the placement of two rather large floodlights on either side of the roof. Photo by Dennis Maag.

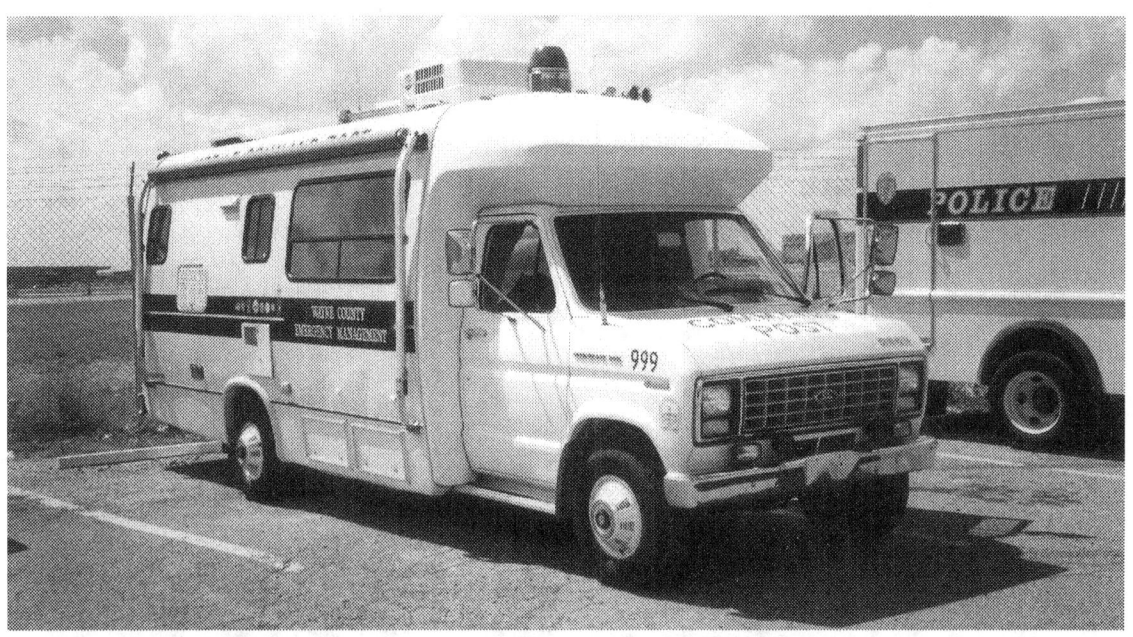

Wayne County Emergency Managment Agency (Michigan) operated this 1987 Ford/Champion motor-home converted into a mobile command post for several years. Note the green strobe light on top of the roof. Photo by Anthony J. Rzucidlo.

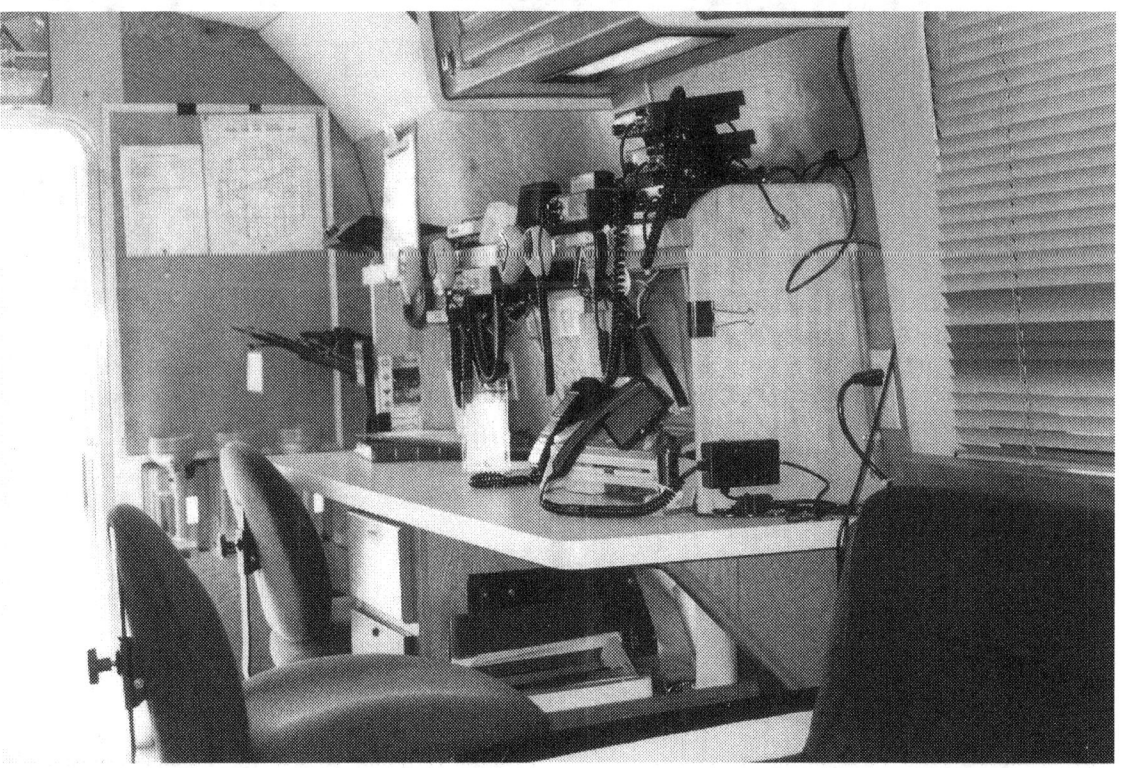

Here is an interior photo of the former Wayne County Emergency Managment's command post showing the numerous radios contained in the communications part of the unit. Photo by Anthony J. Rzucidlo.

Wayne County Emergency Managment replaced it's 15 year old mobile command unit with this 2002 Ford/Winnebago. This unit was a vast improvment from the previous vehicle. This unit when not in operation, is stored at the county sheriff's road patrol office. Photo by Anthony J. Rzucidlo.

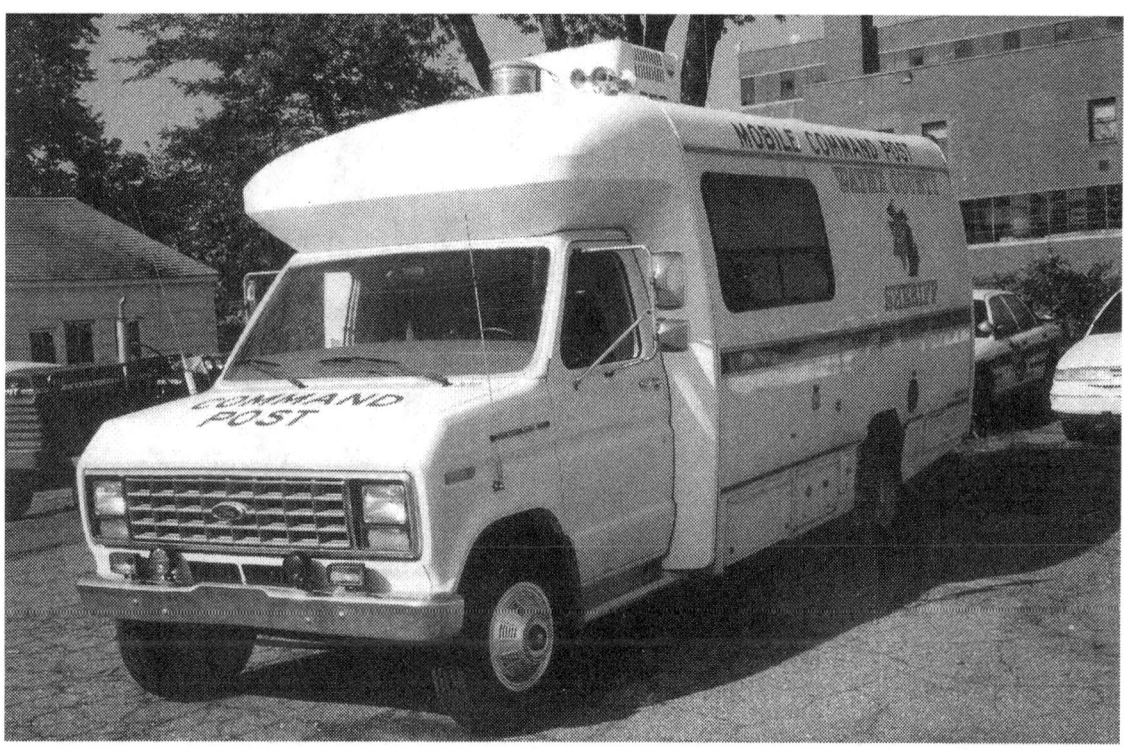

The Wayne County Sheriff Department (Michigan) was given this 1987 Ford/Champion command post upon the county recieving a newer unit for EMA. Note that the green strob command post light is no longer on the vehicle. As of 2006 this vehicle was still in service. Photo by Anthony J. Rzucidlo.

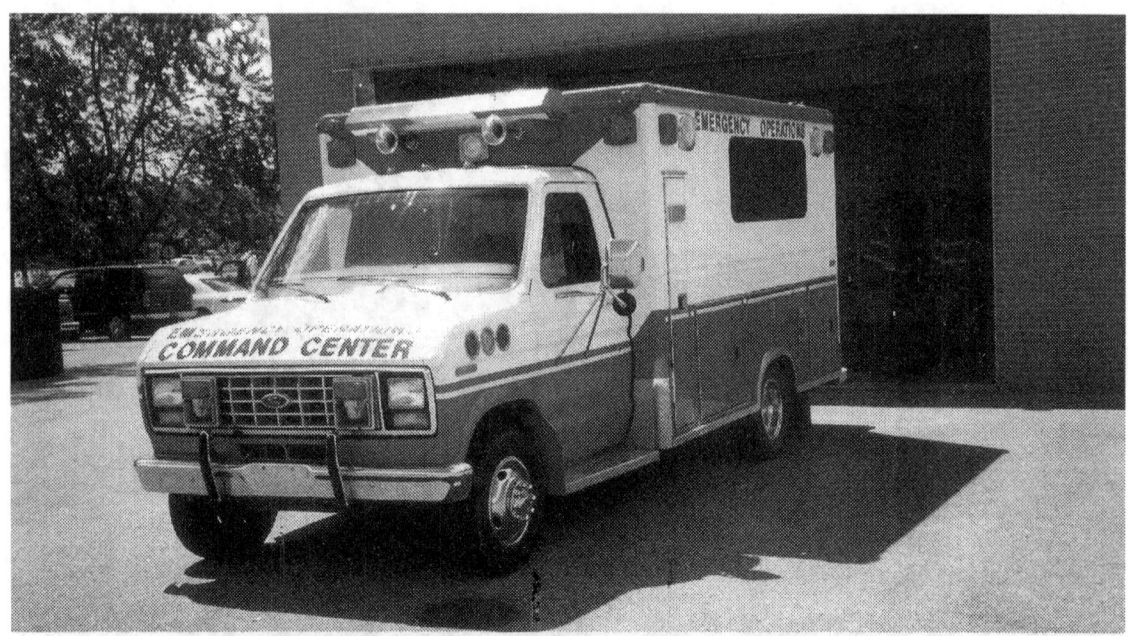

For several years "Battalion Chief's" in the Westland Fire Department (Michigan) used this former 1991 Ford McCoy/Miller ambulance as a mobile command center. They would respond with this unit on calls that they were dispatched to. This unit is no longer in service. Photo by Anthony J. Rzucidlo.

This 1999 Mercedes 112D "Sprinter" with MacNeillie bodywork was assigned to the West Midlands Police (United Kingdom of Great Britian) for use as a motor traffic incident command unit. Note the use of the unique markings on the side of the vehicle and the emergency lights built into the corners of the roof.
Photo by Christopher M. Taylor.

Ypsilanti Police (Michigan) placed into service this 2003 GMC/Utilimaster to serve as the department's mobile command post. The vehicle is painted white over black kind of matching the departments' patrol car fleet. It appears that this unit is equiped with LED emergency lights on the front of this unit as well as in the grill area. Photo by Anthony J. Rzucidlo.

Emergency Response
Support Vehicle Photos

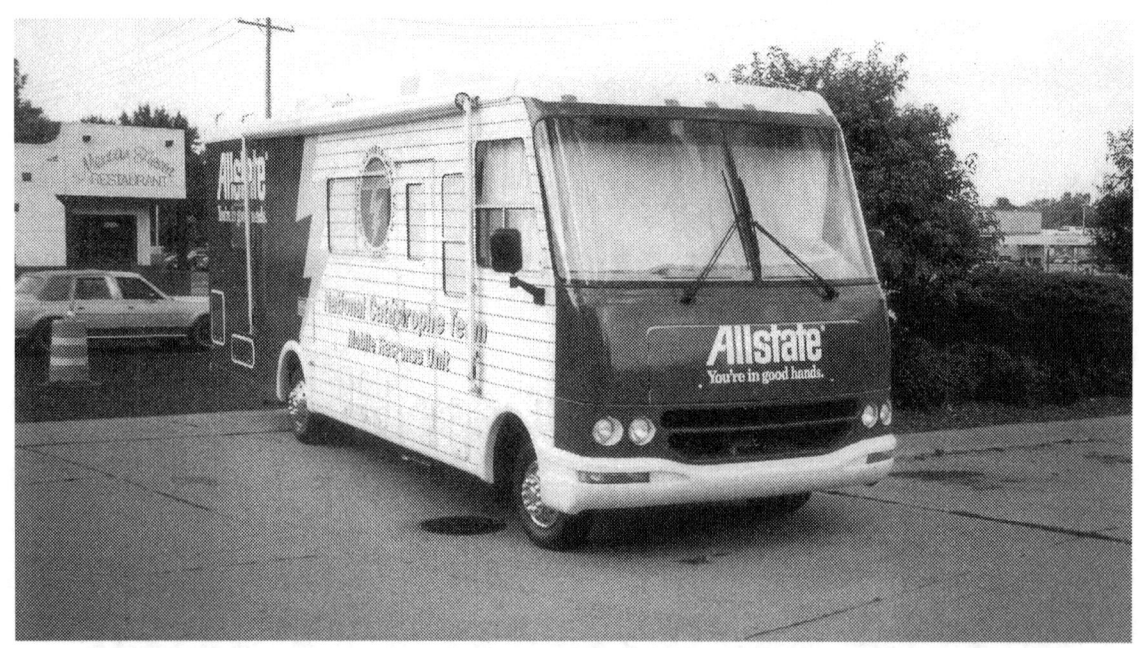

Allstate insurance rolled out this mobile response unit that is part of the "National Catastrophe Team" when severe weather struck the Dearborn (Michigan) area several years ago. Photo by Anthony J. Rzucidlo·

Personnel of the Alton Fire Department (Illinois) operated this 1969 Chevrolet/McCabe-Powers heavy rescue. Originally a Civil Defense truck, thus the blue and white paint scheme was converted into a rescue in later years. In 2003 this unit was replaced by an International/Crimson medium rescue. Photo by Dennis Maag .

**This canteen was operated by the American Red Cross chapter located in Detroit (Michigan) as part of disaster services. Noted on the bottom of this unit indicates that this vehicle was donated by SERVICAR.
Photo by Anthony J. Rzucidlo.**

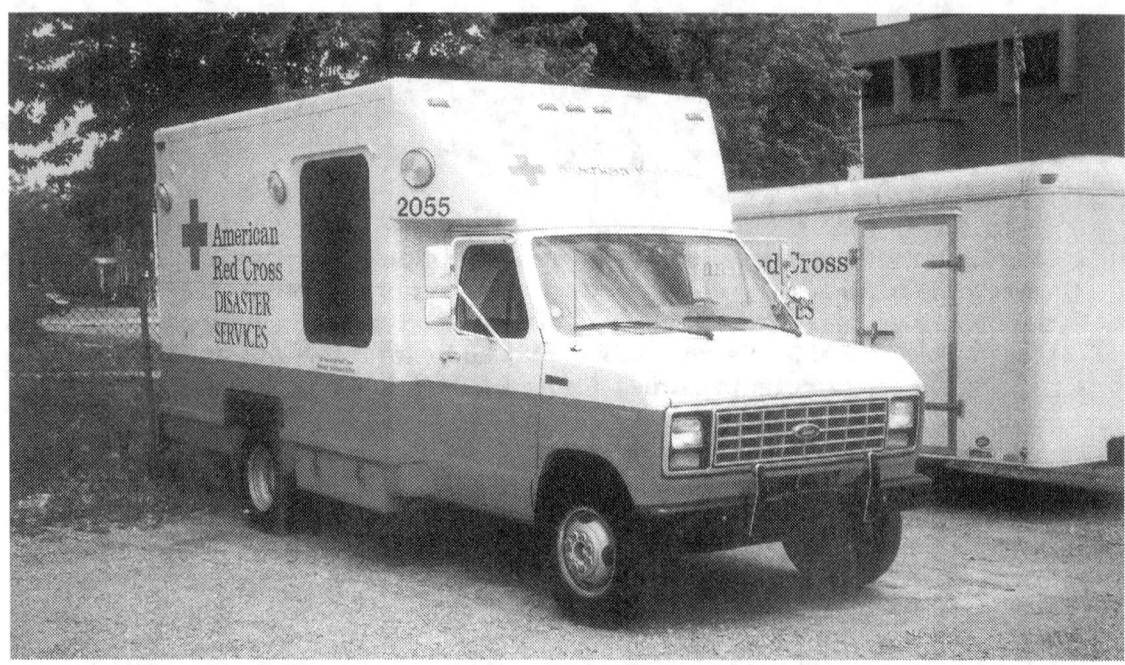

This Ford E350/Utilimaster can be utilized by the American Red Cross in Southeastern Michigan as a canteen unit to feed emergency response personnel at the scene of an incident, or those persons displaced by an emergency type incident. Photo by Anthony J. Rzucidlo.

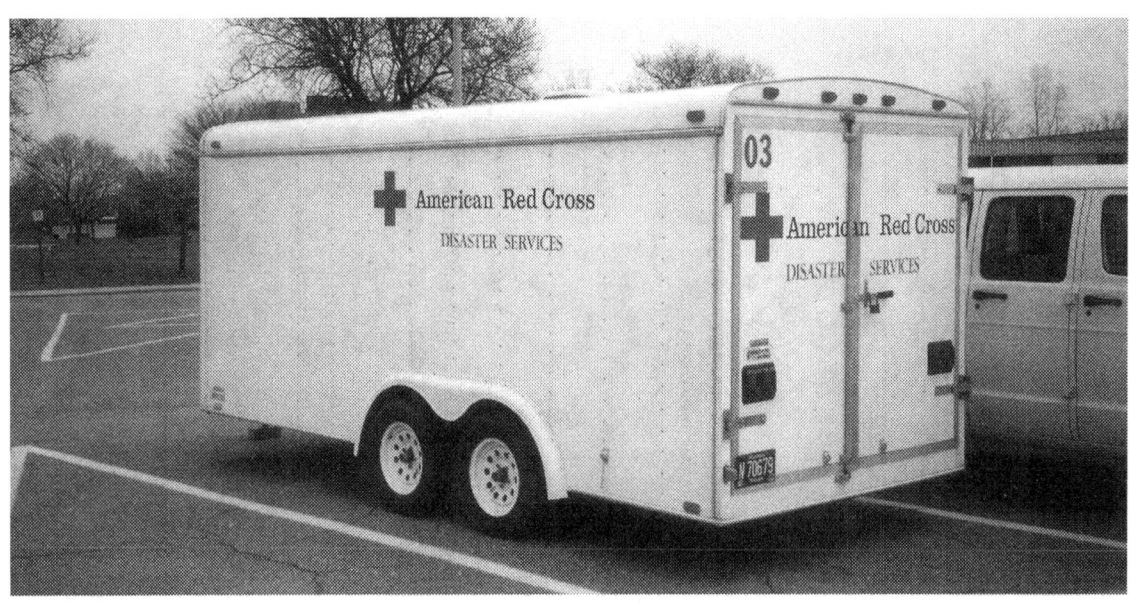

American Red Cross operations in Detroit (Michigan) area have these trailers manufacturered by Southwest located throughout the area. They are equipped with various supplies that include cots, blankets and other personal need items. This trailer was stationed at the Livonia Police Department. Photo by Anthony J. Rzucidlo.

This Ford chassis commuter type bus is utilized by the American Red Cross in the Detroit (Michigan) area as a warming shelter for those persons who may be displaced as a result of fire or other incidents. Photo by Anthony J. Rzucidlo.

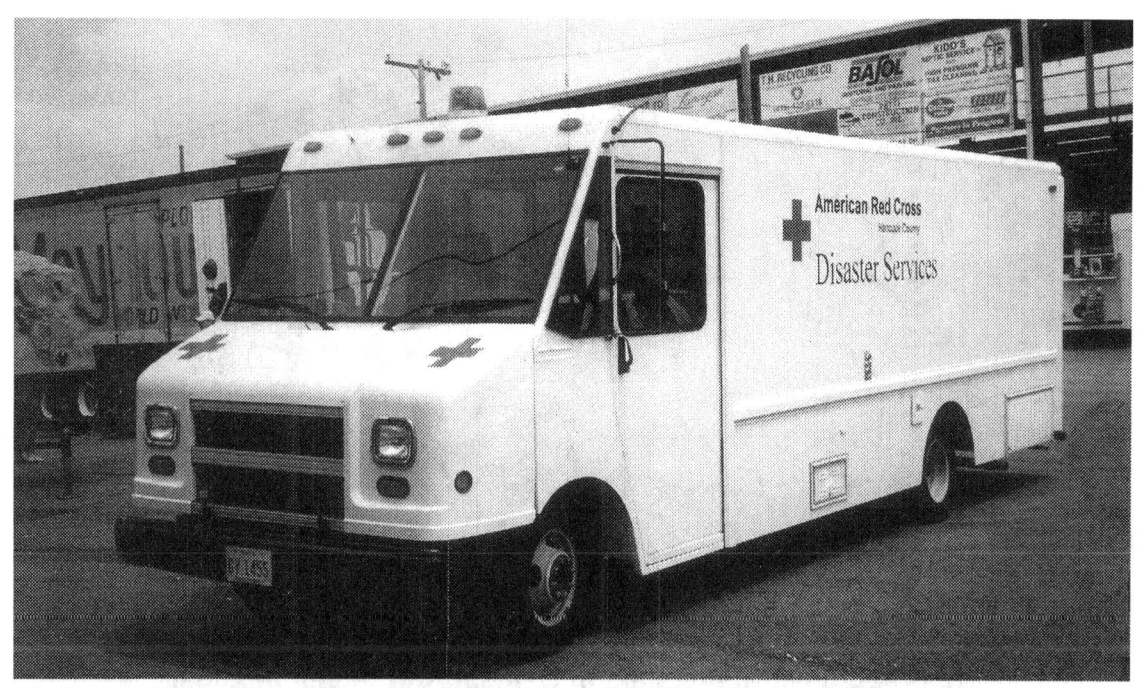

Hancock County in Ohio was the location for this photo of an American Red Cross Ultilimaster step-van. Note that this unit is equipped with a yellow warning flasher on the roof. Photo by Anthony J. Rzucidlo.

Burlington Northern - Sante Fe railroad has trailers such as this trailer made by Featherlite located throughout its system to respond, in the event of a hazardous materials incident. This trailer was on display as part of "Railroad Days" being held in Galesburg (Illinois) in 2000.
Photo by Anthony J. Rzucidlo.

Canadian National railroad a Class 1 carrier in North America utilizes this Chevy van identified as "Mobile 3 Emergency Response" as a hazardous materials unit. Note the raised roof on the van, which will give personnel a little more headroom while working inside the van.
Photo by Anthony J. Rzucidlo.

**This Community Emergency Response Team (CERT) trailer is staffed and
placed into operation at the scene of emergency incidents, by members
of the Charlottsville-Albemarie University of Virginia Community Emergency
Response Team. CERT teams can be found accross the United States.
Photo Courtesy of University of Virginia .**

This former delivery truck a 1988 GMC/Grumman/Olson was obtained by the Coal City Fire Department (Illinois)in the late1990's and converted into a water rescue unit. According to the photographer of this photo, all of the fire apparatus in the department are painted orange. Photo by Dennis Maag.

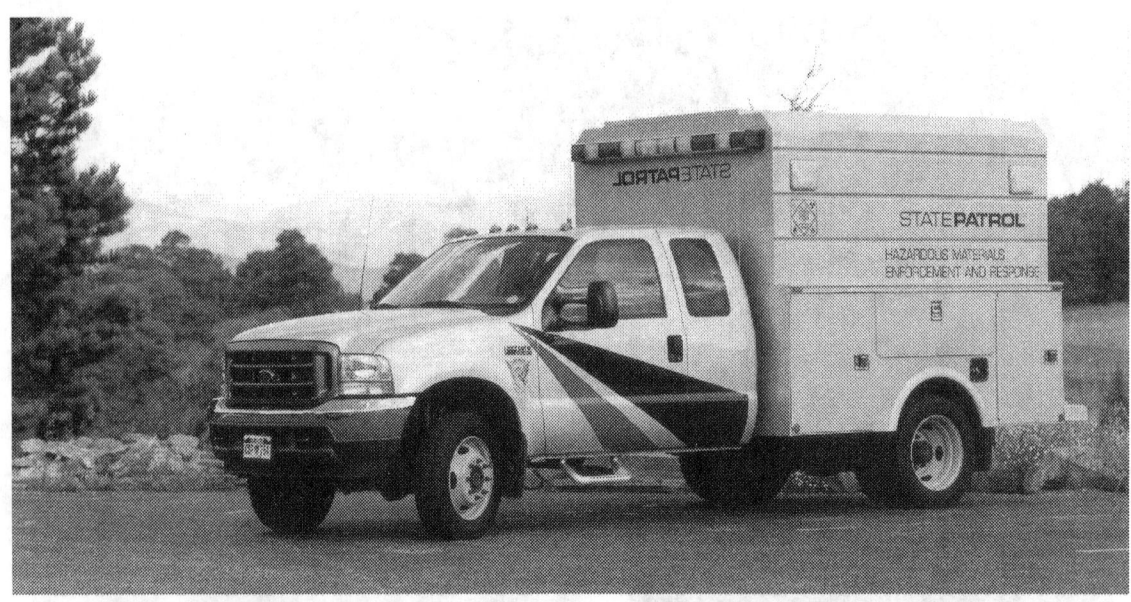

The Colorado State Patrol has 12 of these 2003 Ford chassis trucks located throughout the state, which are broken down into two person response teams. The trucks are done in the same paint scheme and markings as the agencies patrol vehicles. Photo courtesy Colorado State Patrol.

Advanced Containment Systems built this decontamination unit for the Detroit Fire Department (Michigan) in 2000. In adition to this unit, the department has two other pieces of equipment assigned to hazardous materials response. Photo by Anthony J. Rzucidlo.

This rescue squad is operated by the Detroit Fire Department. This unit is a 2003 Spartan/SVI and is equipped with rollup door cabinets and a rear entry walk through body. These vehicles are mainly utilized for rescue type operations at accident or fire scenes. Photo by Anthony J. Rzucidlo.

Detroit Fire Department (Michigan) placed this 2005 Pierce Enforcer into service as "Haz Mat 1." This unit replaced a 1989 Chevrolet/Utilimaster that was the primary response unit. Photo by Sherman Rice.

This 1960 GMC/General was in service in 1991 when this photo was taken as Squad No. 2. When members of the Elmhurst Emergency Services Disaster Agency (Illinois) recieved an Emergency One heavy rescue, this truck was then sent to the town of Winfield (Illinois). It appears, as this truck would have been used as a lighting unit and perhaps had some type of fire fighting capability because of the deck gun mounted on the unit. Photo by Dennis Maag.

The Environmental Protection Agency responded to the metro Detroit (Michigan) area following a hazardous material incident and utilized this Blue Bird bus for air monitoring. Photo by Anthony J. Rzucidlo.

Festus Fire Department (Missouri) operated this 1949 Ford as "Support Unit 33" that was utilized as a generator/air cascade and light unit. The department recieved this unit in the 1950's and as of 1994 was still in service. Photo by Dennis Magg.

In 1996 this Taylor Dunn electric cart was placed into service at the
Ford Research and Engineering Center in Dearborn (Michigan). Security
personnel utilized this unit for fire maintenance work and for special events.
The author of this publication along with another individual from the security
department (Jim Moorer) was responsible for the design and obtaining
this unit. Photo by Anthony J. Rzucidlo.

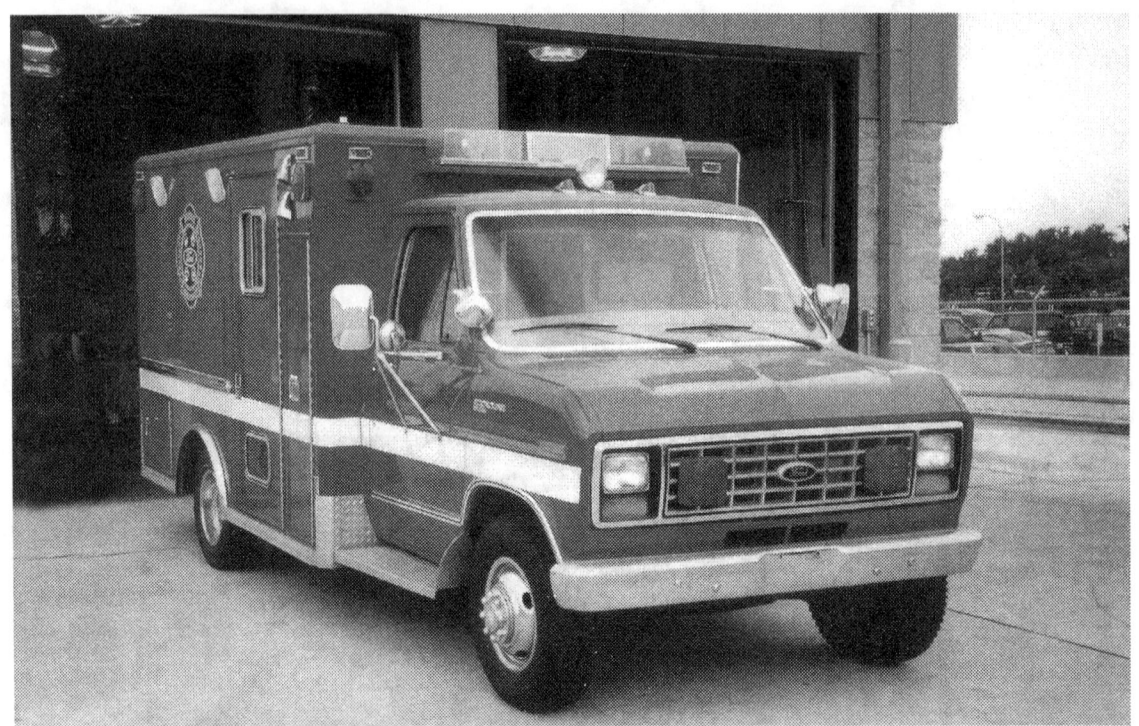

This unit started its career out as an ambulance for the Wayne Fire Department (Michigan). The Ford Motor Company facility in Wayne obtained this 1984 Ford, painted it red and converted it into a rescue unit. Photo by Anthony J. Rzucidlo.

The Ford Motor Company facility located in Windsor (Ontario, Canada) had this Ford van assigned to the sites Emergency Response Team and was utilized for a variety of rescue type incidents. Note the placement of the ladder rack on the roof and the rather large letters spelling out "EMERGENCY" on the side windows. Photo by Anthony J. Rzucidlo.

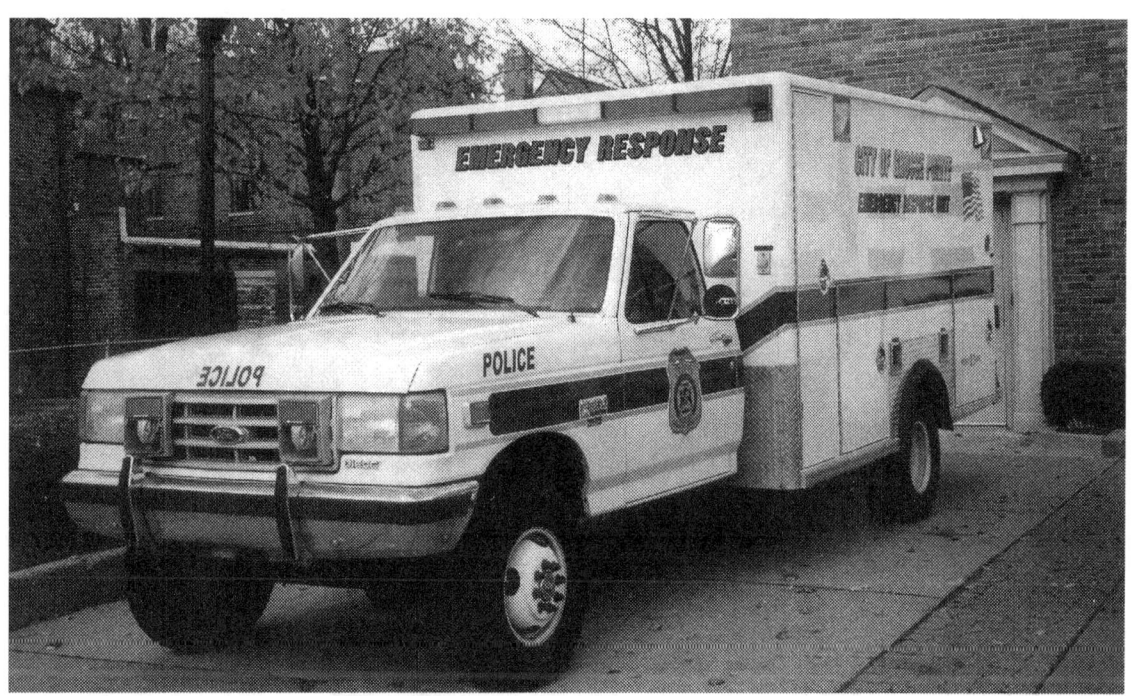

Grosse Pointe City Public Safety Department (Michigan) placed into service this former 1989 Ford/Wheeled Coach U.S. Air Force 4x4 ambulance as a multi purpose unit. The unit is equipped with mass causality items, hazardous materials supplies and can be used as a cooling/warming unit for emergency responders at the scene of an incident. This unit is still in service. Photo by Anthony J. Rzucidlo.

Hamilton County Emergency Managment (Ohio) utilizes this 1970 International/Gerstenlager as a communications center. This unit started out as a bookmobile, and was still in service in 2005 when this photo was submitted. Photo courtesy Hamilton County Emergency Managment.

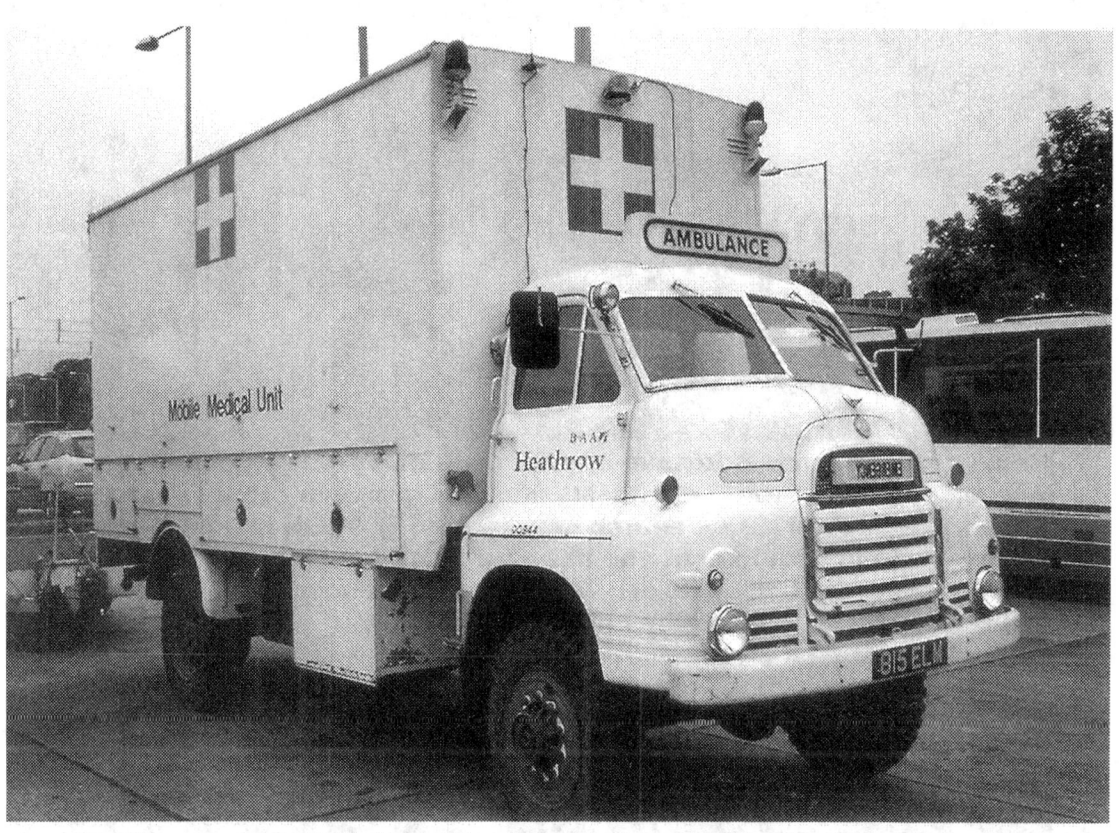

Heathrow airport (United Kingdom of Great Britian) utilizes this 1957 Bedford/Hawson as a mobile medical unit. This unit carries a variety of medical supplies that can be utilized at the scene of a plane crash or any other major incident that could take place at the airport. This unit was reported to still be in service, as late as 2005. Photo by Christopher M. Taylor.

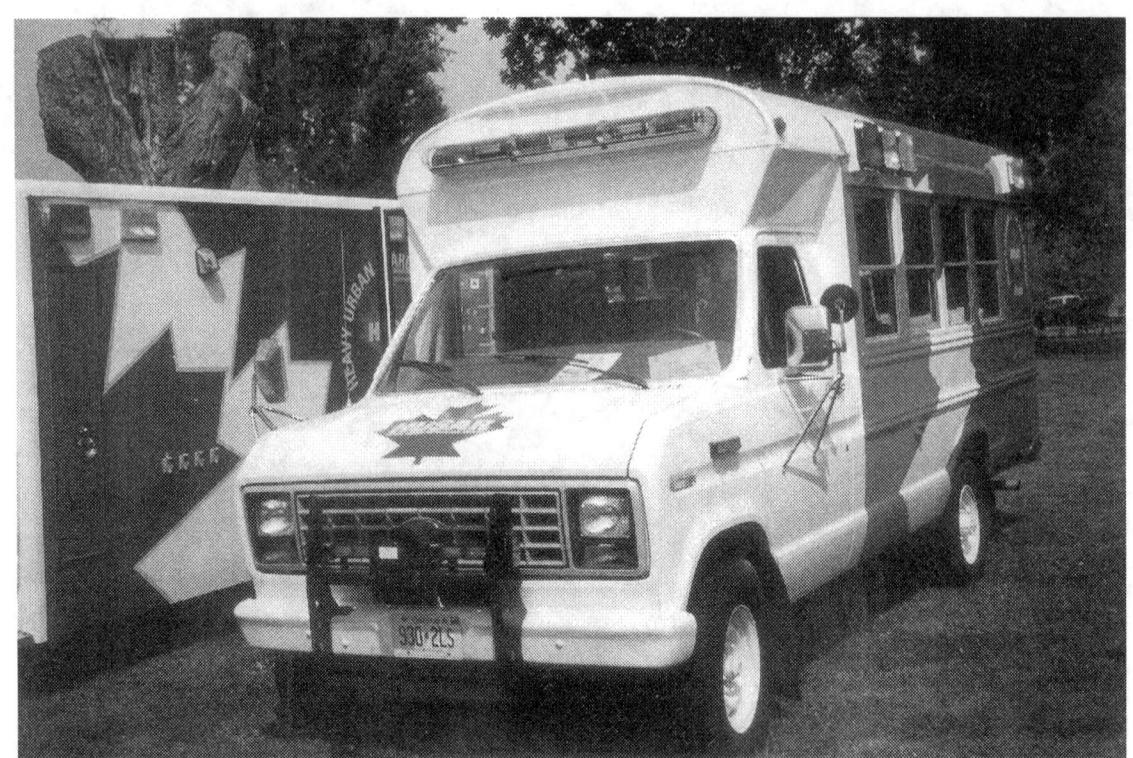

Heavy Urban Search & Rescue operations from Toronto (Canada) have converted this former Ford chassis mini-bus into a unit that is used to transport canine's (K-9) for search operations. This unit is equipped with emergency warning equipment and has an attractive paint scheme of mostly white body with red, white and blue accenting.
Photo by Anthony J. Rzucidlo.

This Volvo flatbed truck is used by Heavy Urban Search & Rescue personnel from Toronto (Canada) to drop off "skids" containing a variety of equipment and supplies. This truck is equipped with a light-bar and red emergency lights attached to the grill. Photo by Anthony J. Rzucidlo.

Huron Valley Ambulance (HVA) located in Michigan utilizes this mini response ambulance at special events to transport ill and injured persons where a normal ambulance would be unable to reach them, because of crowds or narrow access points. Photo by Anthony J. Rzucidlo.

This 2000 Ford F-550 with four wheel drive with an ambulance package is utilized by the Ingham County Sheriff's Department (Michigan) as a heavy rescue unit. This unit is utilized for a number of rescue missions and responds to assist area fire departments when required.
Photo courtesy Ingham County Sheriff.

Metro Lansing Fire Department (Michigan) technical rescue unit radio call sign "Rescue 48" is mounted on a 1990 GMC cube truck with an attached trailer.
Photo by Sherman Rice.

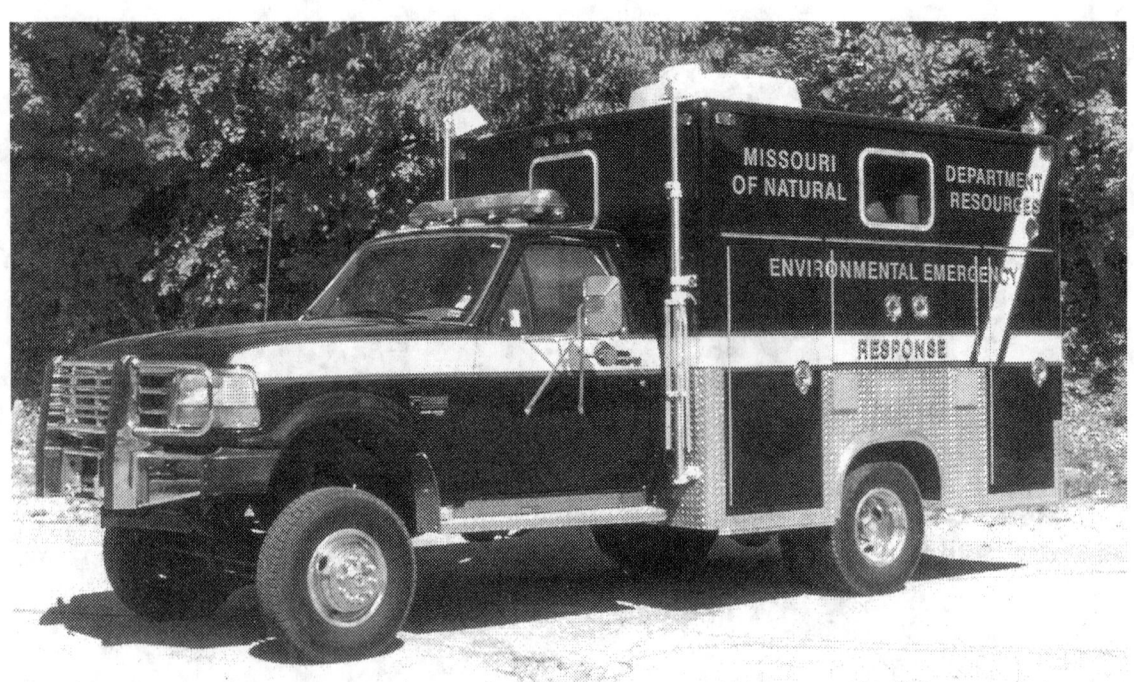

Missouri Department of Natural Resources has six other units like this 1995 Ford F-450 4x4 stationed throught the state. They operate in conjunction with local authorities on any emergency incident that has the possibility of causing damage to the environment. Photo by Dennis Maag.

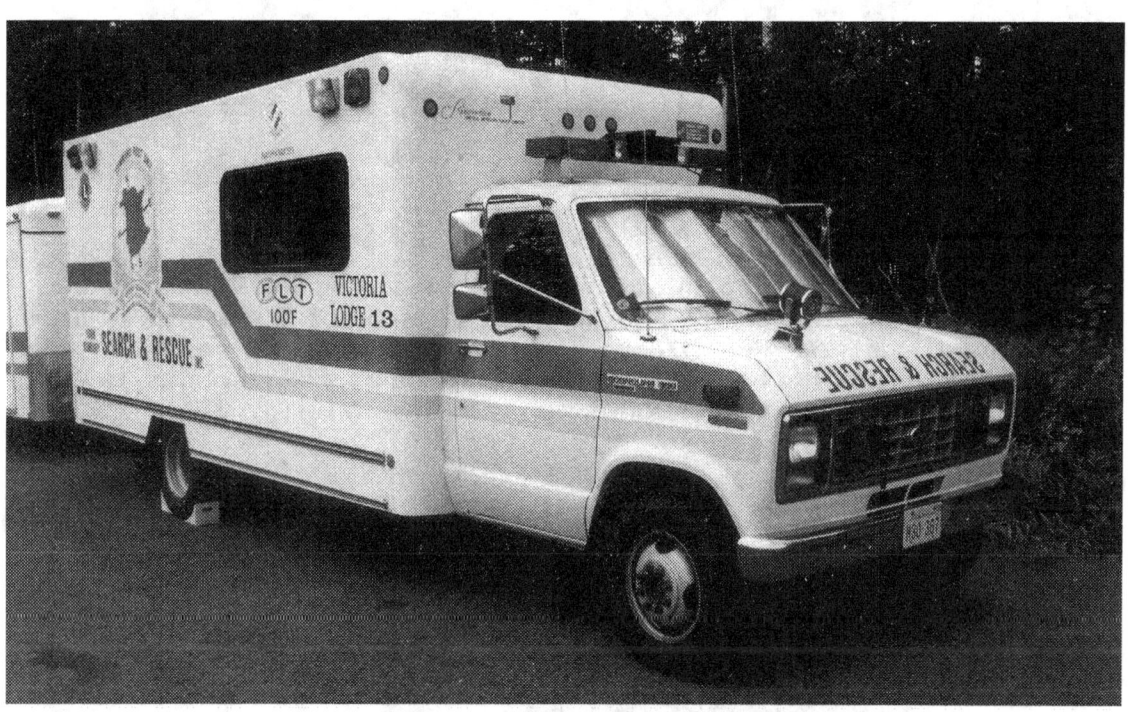

New Brunswick Search & Rescue (Canada) had this 1988 Ford E-350 at it's disposal to carry out it's mission. It appears that some groups may have assisted in donating money towards this unit by the placement of decals on the side of the body. Photo by Christopher M. Taylor.

The Emergency Service unit of the New York City Police Department (New York) operated this 1986 Ford C-900 as a rescue. The unit was painted in the departments' former attractive paint scheme of white over blue. This unit would respond to both police type situations as well as rescue calls. Photo by Christopher M. Taylor.

Emergency Service personnel of the New York City Police utilized this 1987 Grumman for hazardous material support. This unit was also painted white over blue. Photo by Christopher M. Taylor.

Oahu Civil Defense Agency (Hawaii) operated what appears to be a former International Harvester chassis ambulance as some type of response unit. Note the several floodlights on the roof and the additional emergency warning lights also on the roof. Photo from Rzucidlo family collection.

This 1987 Dodge Ram was in service with the Oahu Civil Defense Agency in 2004. Note the CD door decal on the passengers' door. This unit is also equipped with a roof mounted-light bar. Photo by Jeff Spencer.

Oahu Civil Defense (Hawaii) has this 2005 Ford Explorer in it's fleet. On the front fender the following question is asked, " Are You Disaster Prepared?" Also on the side of the vehicle appears the agencies website www.oahucert.com. Also underneath the Civil Defense logo appears Community Emergency Response Team. Photo by Jeff Spencer.

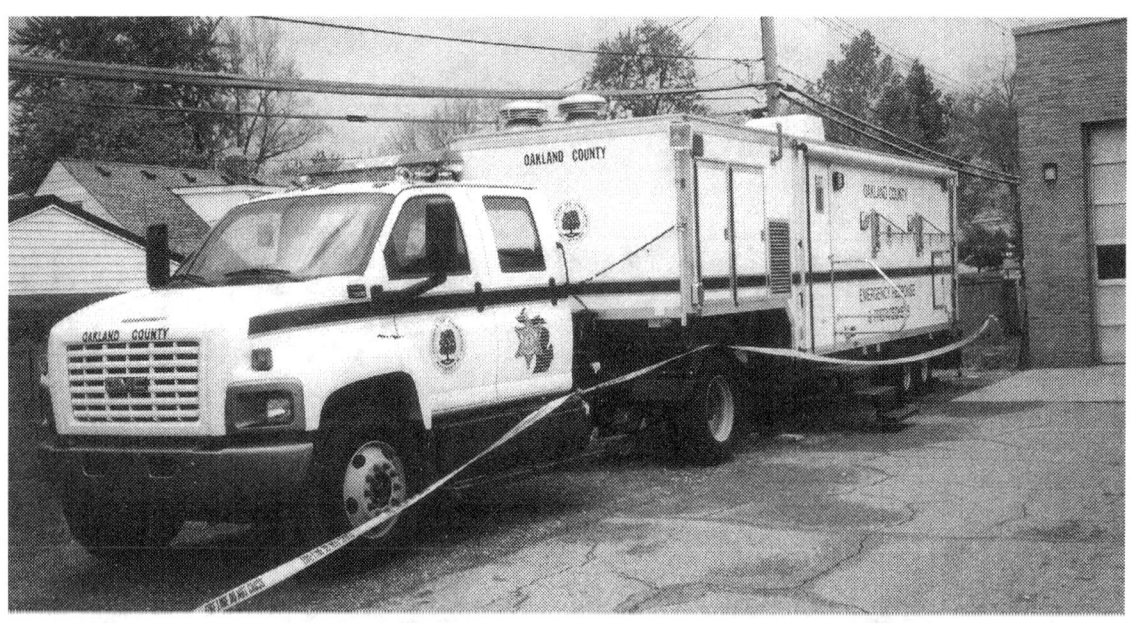

Oakland County Sheriff (Michigan) has this 2003 GMC C7500 tractor that is utilized to pull the attached Advanced Containment Systems decontamination unit, to various hazardous materials incidents throught the county. Photo By Anthony J. Rzucidlo.

Olin Works Fire Department (Illinois) operated this 1969 Ford/Darley pumper originally as a fire engine. In 1992 members of the plant fire department converted it into a hazardous materials unit. This unit was in service until 2002.
Photo by Dennis Maag.

**Port Autority Police Department (PAPD) JFK International Airport (New York)
responded to incidents at the airport with this 1997 Volvo rescue truck.
Note the placement of "Kojak" lights on the extended front bumper for
additional warning requirements.   Photo by Christopher M. Taylor.**

This 1991 Chevrolet step-van is utilized by the Saint Anthony Village Fire Department (Minnesota) as an equipment vehicle. The unit is yellow with the hood painted black. Photo courtesy Saint Anthony Village Fire Dept.

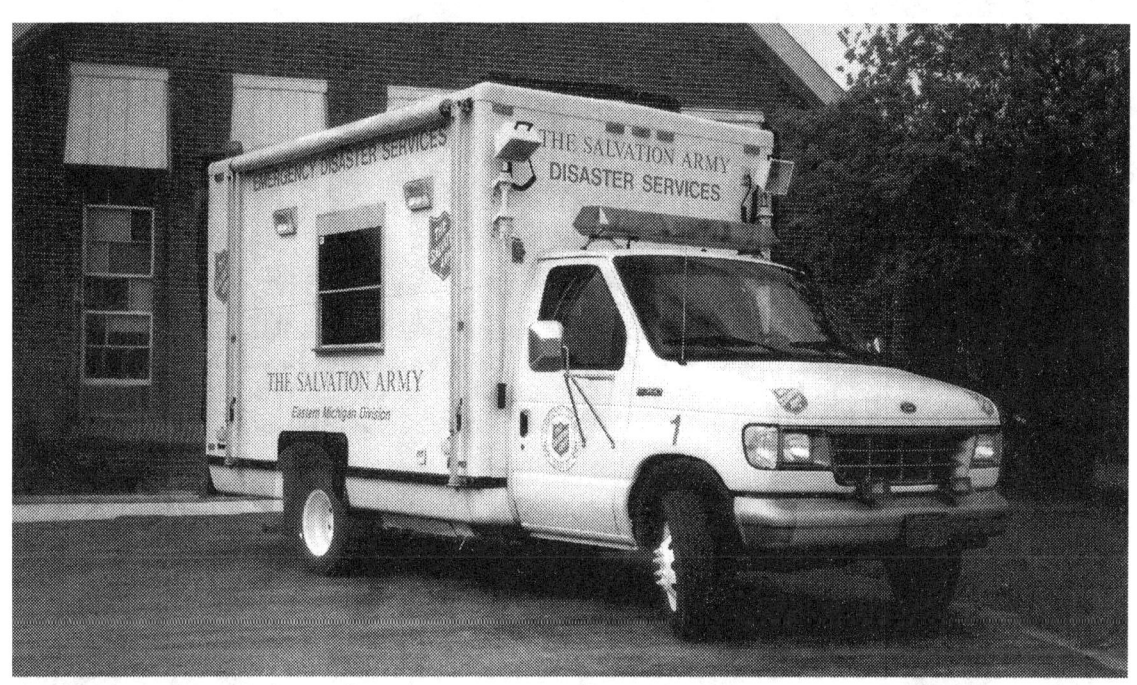

Parked in front of the Salvation Army church in Dearborn Heights (Michigan) is this 1996 Ford chassis canteen unit assigned to the Eastern Michigan Division. Photo by Anthony J. Rzucidlo.

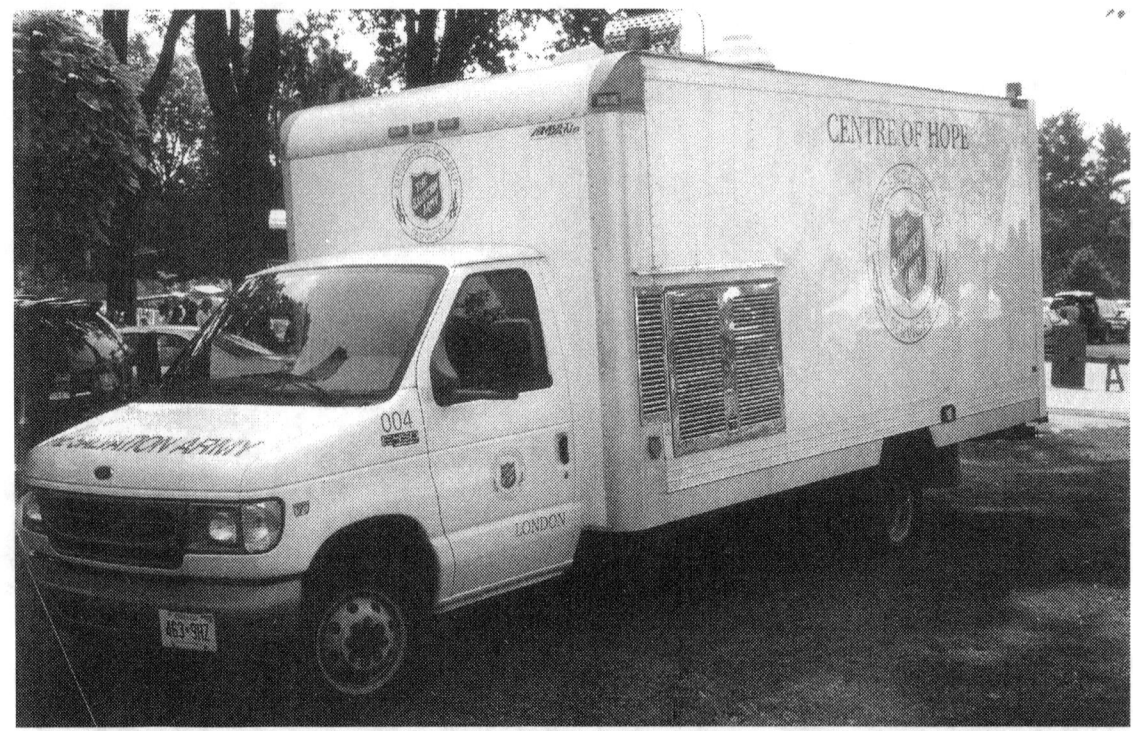

This Salvation Army unit is a Ford E-450/Multivans and is assigned to the office located in London (Ontario, Canada). Note the warning lights located on all four corners of the roof. Photo by Anthony J. Rzucidlo.

Another view of the Salvation Army unit from Canada showing the serving window area and what appears to be a fold down shelf. Also note that this truck is equipped with a fold down awning. Photo by Anthony J. Rzucidlo.

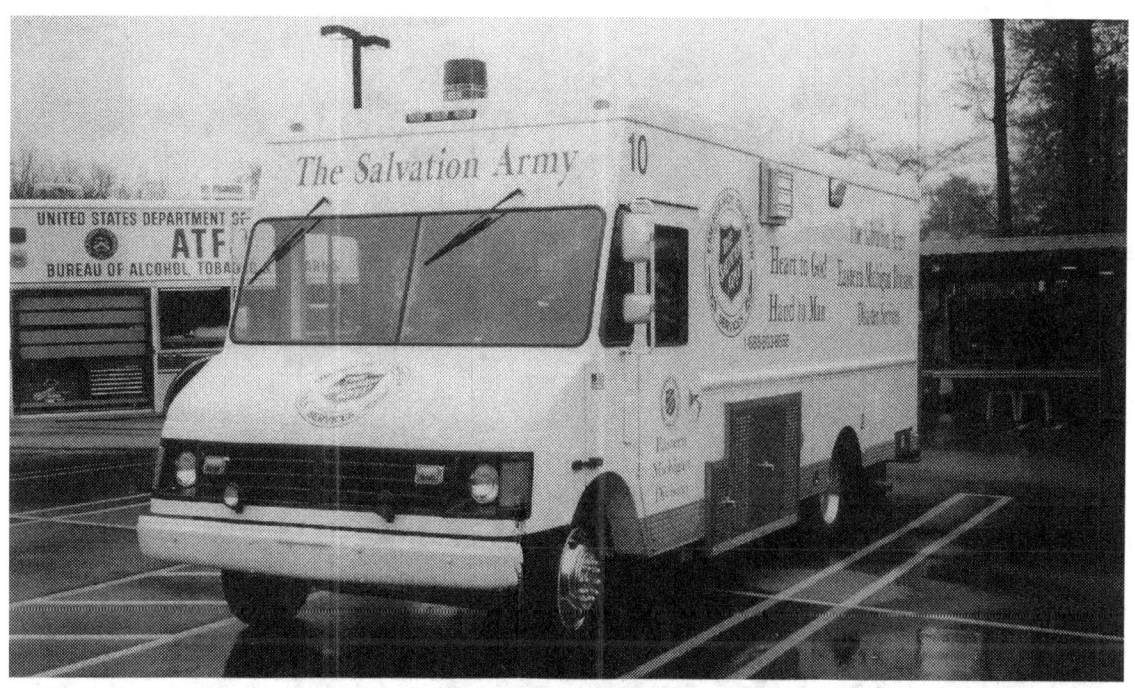

Here is another style of vehicle operated by the Salvation Army in Michigan as a moble canteen unit. This unit is a 2002 Chevrolet/Custom Coach. Note the number of warning lights that this vehicle is equipped with.
Photo by Anthony J. Rzucidlo.

San Diego Fire-Rescue Lifeguard Division (California) utilizes this 1994 International all wheel drive as a rescue unit. "Rescue 44" is a multi purpose emergency respose unit that is equipped to respond to cliff rescues, vertical rescues and swift-water rescues. Photo courtesy San Diego Fire-Rescue Lifeguard Division.

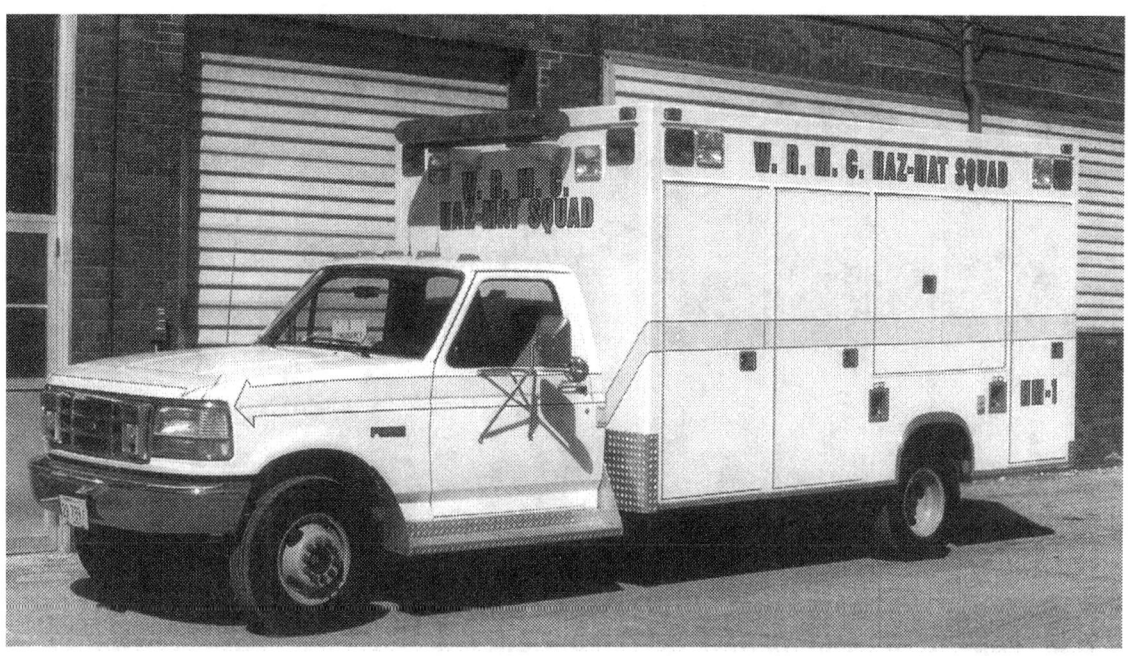

Shell Oil Company - Wood River Manufacturing Center located in Wood River (Illinois) operates this 1992 Ford/Taylor Made as a hazardous materials response unit. The unit is equipped with 12 Level A suits and 20 Level B suits and decontamination equipment. Photo by Dennis Maag.

State Farm insurance like other insurance agencies have a mobile unit that can respond to communities that experience natural disasters. The vehicle is part of the insurance company's catastrophe service. This unit was located in Dearborn (Michigan). Photo by Anthony J. Rzucidlo.

**St. Clair County Sheriff (Michigan) has this GMC chassis mobile
communications unit in it's fleet to respond to any emergency in the county.
Note the air conditioning unit mounted into the body.
Photo by Anthony J. Rzucidlo .**

The goverment of the United Kingdom of Great Britain purchased and placed 14 of these Man/Saxon units throughout the country. They are used for biological, radiological and nuclear decontamination. Photo by Christopher M. Taylor.

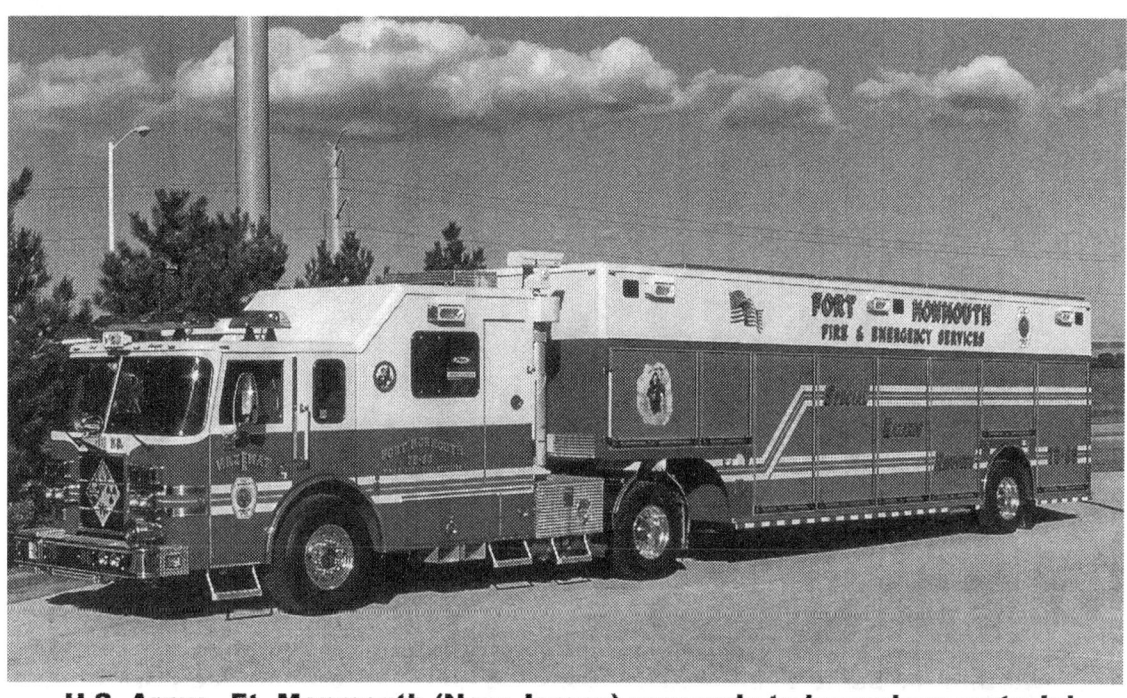

U.S. Army - Ft. Monmouth (New Jersey) responds to hazardous material incidents on the base with this tractor/trailer unit. The tractor is a 2004 Pierce Lance. It appears that command operations associated with the incident could be centered in the rear portion of the tractor unit. Photo by Dennis Maag.

Universal Ambulance (Michigan) operates this 1984 Chevrolet step van with attached trailer as a hazardous materials/decontamination unit. The van is painted in the same paint scheme as the company's ambulances' which is white over orange. Photo by Sherman Rice.

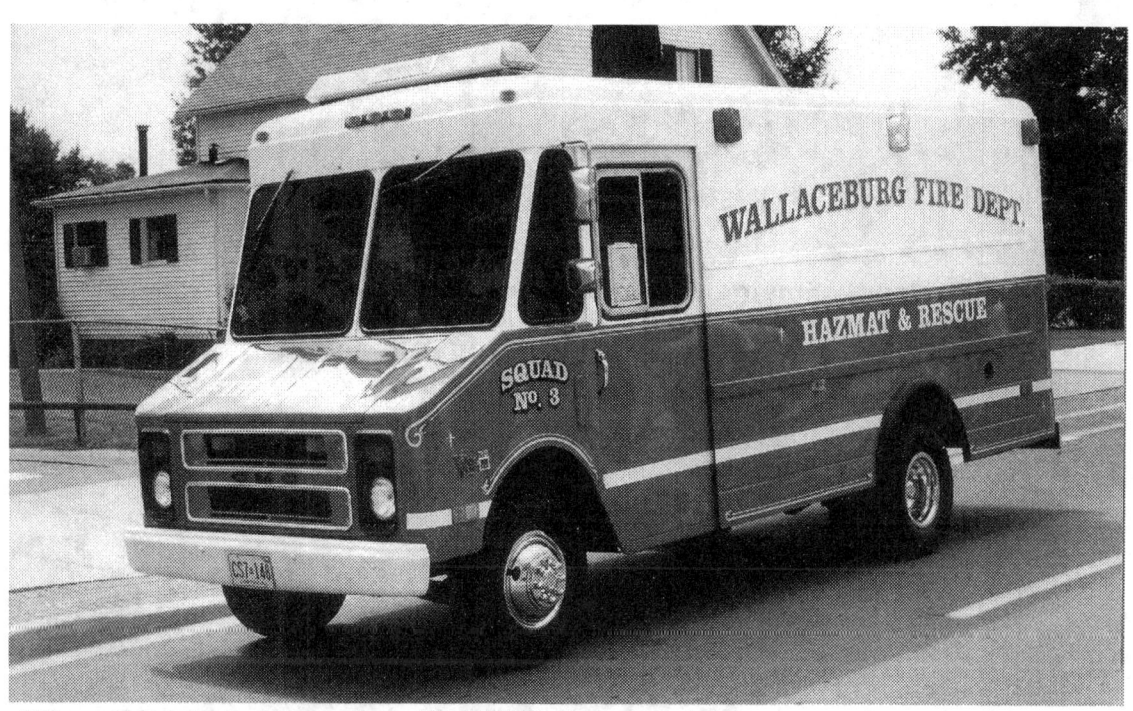

When this photo was taken this GMC chassis step van was in operation with
the Wallaceburgh Fire Department (Ontario, Canada) and was set-up to
respond to hazardous materials incidents or rescue calls. This unit was
painted white over red with gold lettering, which made for an attractive
piece of apparatus. This unit was still in service in 2005.
Photo by Anthony J. Rzucidlo.

Wayne County Emergency Managment (Michigan) has this former Chevrolet/Horton ambulance in it's fleet and is utilized for the transportaion of equipment or personnel. This unit was still in service as of 2005. Photo by Anthony J. Rzucidlo.

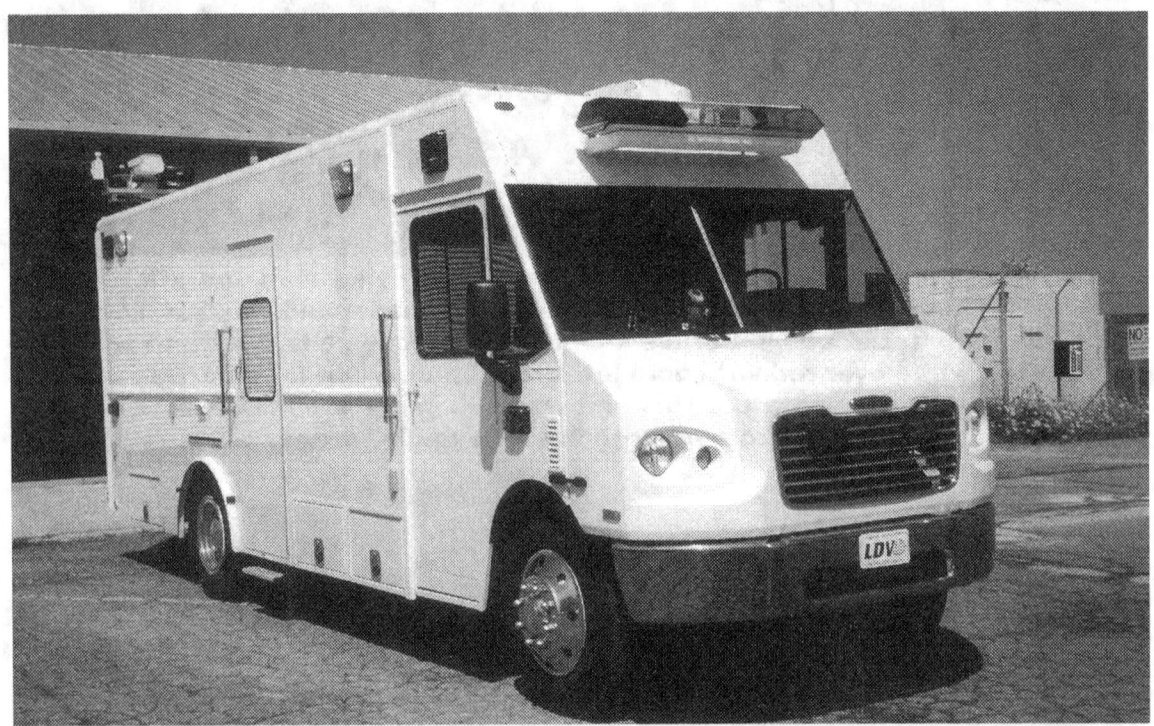

This 2003 Frieghtliner/LDV is utilized as a support vehicle for Wayne County Emergency Managment's (Michigan) mobile command center. The vehicle is painted white and at the time of this photo had no markings applied to it. Photo by Anthony J. Rzucidlo.

Western Wayne County Hazardous Materials Team (Michigan) utilized this GMC truck with attached trailer for hazardous materials incidents in that part of the county. Photo by Kimberly D. Rzucidlo.

Western Wayne County (Michigan) Fire Mutual Aid Association operates this Ford F-450 with an attached trailer for hazardous material responses. Photo by Anthony J. Rzucidlo.

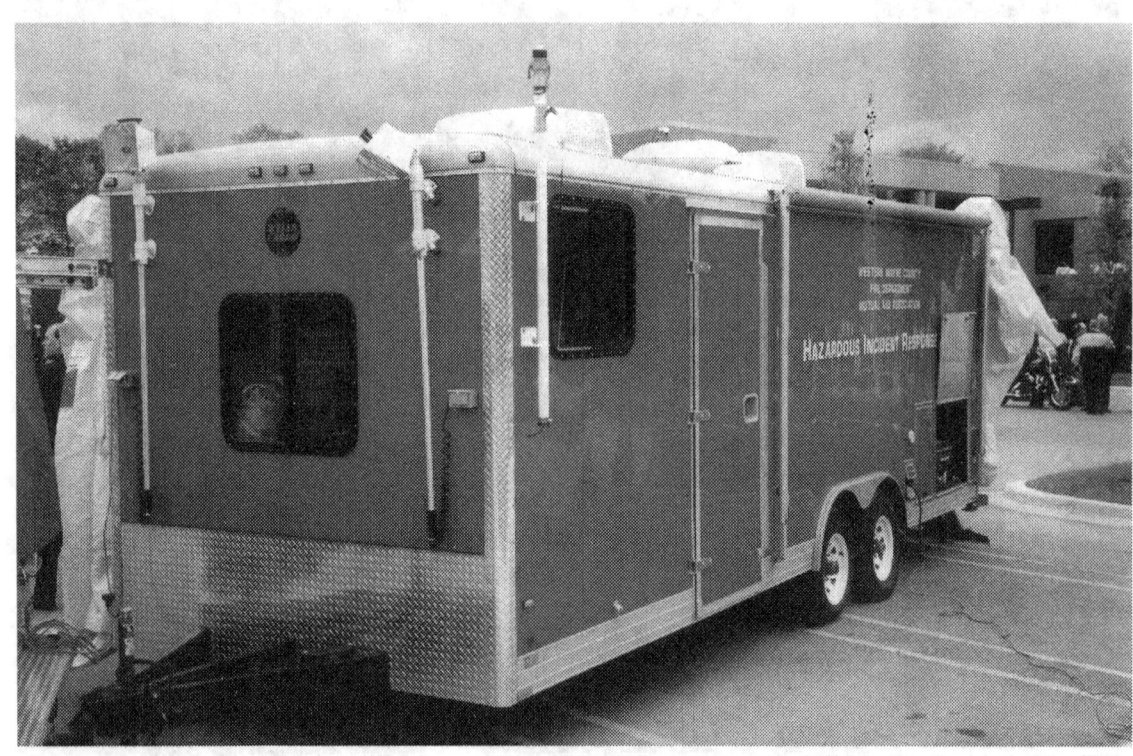

This Wells Cargo trailer is also utilized in the response of hazardous material incidents by members of the Western Wayne County Mutual Aid Association. Photo by Anthony J. Rzucidlo.

West Routt Fire District - Hayden County (Colorado) operated this 1982 Peterbilt 6x6/Emergency One as a rescue unit. If nessasary this vehicle can serve as a command post and a comunications center on major alarms. Photo by Dennis Maag.

Photographs

Photographs that appear in this publication were taken in the following states and or countries: California, Canada, Colorado, Florida, Hawaii, Illinois, Massachusetts, Michigan, Minnesota, Missouri, Nevada, New York, North Carolina, Ohio, Texas, United Kingdom of Great Britain and Virginia.

Photographs that appear in this publication were either taken by, or submitted from the following individuals or organizations. I wish to thank them for their contributions. Without them, this publication would not have been possible.

Individuals
Dennis Maag
Christopher M. Taylor
Sherman Rice
Rzucidlo Family
Jeff Spencer

Organizations
Ann Arbor Office of Emergency Management, Michigan
Ajax Fire & Emergency Services, Canada
Alaska Railroad Corporation
British Transport Police, United Kingdom of Great Britain
California Emergency Mobile Patrol Search & Rescue
Charlottesville - Albemarle Community Emergency Response Team UVA
Charlottesville Fire Department, Virginia
Coiler County Emergency Management, Florida
Colorado State Patrol
Hamilton County Emergency Management, Ohio
Ingham County Sheriff's Office, Michigan
Ithaca Police Department, New York
Livermore Police Department, California
Lyons Police Department, Illinois
Massachusetts Bay Transit Authority Police
Nevada Department of Public Safety
New Rochelle Police Department, New York
Orlando Police Department, Florida
Pasadena Police Department, California
Rowan County Telecommunications, North Carolina
Saint Anthony Village Fire Department, Minnesota
San Diego Fire-Rescue Dept./Lifeguard Division, California

About the Author

After becoming an Eagle Scout, I was a member of the Dearborn Heights Police Explorer Post 1809 for three years. I worked for the Dearborn Heights Police Department in a civilian capacity (clerk/dispatcher) for 13 months after high school.

I am a third generation Ford Motor Company employee. The Ford Motor Company has employed me for 32 years, all of which have been spent in security. I started as a guard at the Research & Engineering Center in Dearborn, Michigan. I was promoted to a clerk's position and was assigned to the Ford Rouge Center Fire Services and Rouge Center Security. For a few months in 1989 I was assigned to the Corporate Communications Center (dispatch) in Dearborn as one of the original members of this department. On July 16, 1989 I was promoted to a security supervisor at the R&E Center where I served as a shift operations supervisor, then fire evacuation and training supervisor. Shortly after that, I was placed in charge of special events, emergency preparedness and training while still assigned to the Research & Engineering Center.

In late 2000 I was assigned to Corporate Security where I was the supervisor responsible for emergency preparedness and special events. In early 2001, I returned back to the Rouge where once again I was a shift supervisor. In 2003, I was re-assigned back to being responsible for training at the Rouge Center for security personnel.

During the course of 2005, the company outsourced uniformed security/fire services that were provided by Ford's in-house security organization in North America. I still remained a Ford employee in the Ford security department, and as a result of the changes, my duties and responsibilities were changed. I was re-assigned to the Corporate Security/Fire Department at World Headquarters in Dearborn, Michigan once again.

In 1997 the Ford Motor Company Research & Engineering Center and the City of Dearborn took part in a major disaster drill exercise that was referred to as EX-97. I served as the Incident Commander for the Ford Motor Company during the course of the exercise and took part in "Unified Command" along with the Dearborn fire and police department Incident Commanders. The exercise/drill was one of the largest in the state and the county at that time and involved 25,000 people.

The exercise centered on a weather-related event (tornado) and associated incidents from the tornado that resulted in fires, hazardous material incidents, as well as injuries. The sheltering of employees who were displaced from this incident was also part of the exercise, as well as business resumption.

With my involvement as part of a team that was involved with the development of a new emergency preparedness plan for the R&E Center, we received a Customer Driven Quality Award for our efforts in 1997.

Over the course of my 32 years I have had the opportunity to respond to several emergency incidents and served as the Incident Commander. I also responded to the Ford Rouge powerhouse explosion that occurred on February 1, 1999. I was responsible for the operation of the security command post.

I have two college degrees, Associates in Industrial Security and a Bachelor's in Industrial/Institutional Security with a minor in Law Enforcement.

I served in the capacity as chairperson for the Public Safety Committee for the Dearborn Heights Strategic Planning Task Force. Our recommendations resulted in a new fire headquarters in 1999 and a new Justice Center (combined court/police facility) in 2003. I attended and completed the Dearborn Heights Citizens Police Academy. In addition, my wife and I completed training and have been certified as members of the Dearborn Heights Community Emergency Response Team (CERT). In fact I am one of the instructors in the area of the CERT Organization/ Incident Command. Also I am one to the two-team leaders for our CERT program.

Besides taking photos of emergency response vehicles, which I feature, on my own web site at www.geocities.com/ajrcommander, I have a collection of Ford security; police explorer and railroad police shoulder patches. I am also interested in railroads, and I am the President/CEO of the Michigan Railroad Club, Inc. In addition I belong to a local fire buff club (Box 42).

Prior to this book, I have written one other book and that was on the occasion of the 50th anniversary of the Michigan Railroad Club, Inc.